Science Education and Student Diversity
Synthesis and Research Agenda

The achievement gaps in science and the underrepresentation of minorities in science-related fields have long been a concern of the nation. This book examines the roots of this problem by providing a comprehensive, "state-of-the-field" synthesis and analysis of current research on science education for minority students. Research from a range of theoretical and methodological perspectives is brought to bear on the question of how and why our nation's schools have failed to provide equitable learning opportunities in science education to all students. From this wealth of investigative data, the authors propose a research agenda for the field, identifying strengths and weaknesses in the literature to date as well as the most urgent priorities for those committed to the goals of equity and excellence in science education.

Okhee Lee is a professor in the School of Education at the University of Miami in Florida. She completed her Ph.D. in educational psychology with a focus on science education from Michigan State University in 1989. Her research areas include science education, language and culture, and teacher education. She was awarded a 1993–1995 National Academy of Education/Spencer Postdoctoral Fellowship, and she was a 1996–1997 Fellow at the National Institute for Science Education at the University of Wisconsin-Madison. She received the Distinguished Career Award from the American Educational Research Association (AERA) Standing Committee for Scholars of Color in Education in 2003. She has directed numerous research and teacher enhancement projects funded by the National Science Foundation, U.S. Department of Education, Spencer Foundation, and Florida Department of Education. Her research has appeared in prominent journals in education, including the *American Educational Research Journal, Educational Researcher, Review of Educational Research, Review of Research in Education, Teachers College Record, Journal of Research in Science Teaching, Science Education,* and *Bilingual Research Journal.*

Aurolyn Luykx is a joint associate professor of anthropology and teacher education at the University of Texas at El Paso. She completed her Ph.D. in linguistic and educational anthropology from the University of Texas at Austin in 1993 and was later awarded "best dissertation of the year" from the Council on Anthropology and Education. She is the author of *The Citizen Factory: Schooling and Cultural Production in Bolivia* (1999), which has been used in numerous college courses in both education and Andean studies. She spent several years in Bolivia working on various aspects of that country's nationwide educational reform, and she later became one of the founding faculty of the Programa de Formación en Educación Intercultural Bilingüe para los Paises Andinos (PROEIB Andes), an international master's program for indigenous educators throughout the Andean region. During this time she also received a National Academy of Education/Spencer Postdoctoral Fellowship for study of the use of indigenous languages in higher education. In 2001 she relocated to Miami as a researcher on Okhee Lee's project, Science for All, funded by the National Science Foundation. Together, Drs. Luykx and Lee have published numerous articles on science education for students from linguistically and culturally diverse backgrounds. Dr. Luykx's work has been published in the *Journal of Latin American Anthropology, International Journal of the Sociology of Language, American Educational Research Journal, Teachers College Record,* and *Journal of Research in Science Teaching.*

Science Education and Student Diversity

Synthesis and Research Agenda

OKHEE LEE
University of Miami

AUROLYN LUYKX
University of Texas at El Paso

CAMBRIDGE UNIVERSITY PRESS
Cambridge, New York, Melbourne, Madrid, Cape Town, Singapore, São Paulo

Cambridge University Press
40 West 20th Street, New York, NY 10011-4211, USA

www.cambridge.org
Information on this title: www.cambridge.org/9780521859615

First published 2006

Printed in the United States of America

A catalog record for this publication is available from the British Library.

Library of Congress Cataloging in Publication Data

Lee, Okhee, 1959–
Science education and student diversity : synthesis and research agenda/
Okhee Lee, Aurolyn Luykx.
    p.   cm.
Includes bibliographical references.
ISBN 0-521-85961-1 (hardback) – ISBN 0-521-67687-8 (pbk.)
1. Science – Study and teaching – United States.   2. Multicultural education – United States.
I. Luykx, Aurolyn, 1963–   II. Title.
LB1585.3.L44   2006
507′.1073 – dc22   2006009562

ISBN-13   978-0-521-85961-5 hardback
ISBN-10   0-521-85961-1 hardback

ISBN-13   978-0-521-67687-8 paperback
ISBN-10   0-521-67687-8 paperback

Contents

Foreword

In *Science Education and Student Diversity*, Okhee Lee and Aurolyn Luykx have achieved a comprehensive and authoritative treatment of all aspects of the topic: policy, conceptual frameworks, student characteristics, instruction, curricula, assessment, teacher preparation and professional development, school organization, and the relationships of science education to families, home environments, and communities of diverse students. It is difficult to imagine any serious educator of our time who will not be grateful for a reading of this book. The authors have gathered all the facts, given us a calm and convincing critique of our state of knowledge and practice, and drawn wise conclusions as to where and how our knowledge can further grow.

This book takes on even greater importance from the context of its creation. The authors headed a team of scholars from several research institutions, collaborating through programs of CREDE, the Center for Research on Education, Diversity & Excellence, now located at the University of California, Berkeley. From 1996 through 2004, CREDE was the national research center of the U.S. Department of Education, concerned with research and development of effective educational programs for students of diverse languages, races, cultures, economic strata, and geographies – those students placed at risk of failure in schools by traditional programs designed for mainstream society. CREDE's 40 research projects (and 80 affiliated researchers) spanned the United States, from Hawaii to Florida, from Alaska to Providence, studying students of every major linguistic and cultural group. Our purpose and our achievement was to understand clearly issues of local and specific variation and to discern the underlying principles that can guide effective program design.

In the last two years of our national center work, synthesis of research results was a central focus. The authors of this volume led CREDE's *synthesis team* on Science Education and Diversity. They joined other sister synthesis teams,[1] each focused on specific topical domains of diversity and education.[2] Their purpose was to assemble and synthesize the domain's research evidence and to present it with two foci: what we know now and what we need to know next, so that clarification of the research literature can guide future inquiry.

The universe of knowledge addressed by the synthesis scholars was international. Though the preponderance of published research comes from the United States, issues of societal diversity are now global. Researchers in many nations are informing one another across borders and populations. The corpus of research reports is also heavily weighted with authors affiliated with CREDE. That "accident" is certainly due to the excellence of their work, but also to the good fortune of the generous funding available to CREDE. We were blessed with disproportionate resources as compared with our other colleagues. Because education-and-diversity research was for decades of little interest to mainstream educators and researchers, funding was meager and interest in the topic was slight. CREDE existed in that brief historical period when diversity research bobbed up in national policy concerns. There is no longer a national research center concerned with diversity, even though diversity of our population continues to grow, and the achievement gap between mainstream students and those placed at risk continues.

To assure that all pertinent research was considered in our syntheses, each team was balanced in two dimensions: CREDE- and non-CREDE–affiliated scholars, and diversity and mainstream scholars. The latter balancing was strategic. Since diversity research began in the 1960s, little attention has been paid by mainstream researchers, even in the same domain; and insufficient attention to mainstream research has been paid by diversity researchers. As with two circulating pools in the same lake, little mutual influence was exerted. Our synthesis teams were (metaphorically) locked in the same room for two years and not let out until they had synthesized. The results have been a uniquely rich set of reports.

Our hope is that this volume (and its sister reports) will be of interest to all researchers and policymakers in each domain. In the last six years, educational policy has heavily emphasized research-based practice. All readers of this book surely welcome that emphasis, while regretting that research on culturally and linguistically diverse students is rarely considered in current federal interpretations. In the resulting one-size-fits-all policy climate,

---

[1] Professor Yolanda Padron at the University of Houston provided the organization and coordination of the synthesis teams.
[2] The synthesis teams and their reports are discussed later in this Foreword. During the time of our planning, the synthesis of research in mathematics and diversity education was being organized separately by NCISLA, the National Center for Improving Student Learning in Mathematics and Science (University of Wisconsin, Madison).

our goal is to produce an appreciation for how research *on and for diverse students* should be the basis for educational practice in a diverse society.

Each synthesis team found a unique state of knowledge in its domain. The authors of the current volume present a rich bibliography of research, conducted by many methods and many designs, with a complex field of findings that illuminate a set of still-to-be-investigated important hypotheses. By contrast, the team synthesizing issues of Professional Development and Diversity found a wealth of policy speculation and few systematic studies of variations in preparing teachers for diverse classrooms. Not every synthesis report is of book length, but in each instance the synthesis work clarifies and charts a future research agenda.

Likewise, each synthesis team chose a somewhat different filter for inclusion. Overall, our syntheses program adopted one general inclusion rule: each team discusses the best available research in its domain. The inclusion rules are important to understand in the context of current research-design dialogue. Federal policy's recent emphasis on the Randomized Field Trial (RFT) design was an inevitable corrective to a declining discipline in educational research. Perhaps the RFT advocates are moderating their initial rhetorical excesses ("There is RFT and all else is myth"), but in any event a wiser and more balanced view of design proprieties will emerge, so that different methods and designs are understood as appropriate for different developmental stages of a domain of inquiry. In that spirit, each team adopted a different filter of inclusion, depending on the maturity of the domain. This strategy illuminates the future research agenda, and indeed suggests the methods appropriate to forward the developmental progress.

In this volume, Lee and Luykx recognize that science education and diversity is a relatively new field of inquiry, coming into focus only in the 1990s. Inclusiveness in methods of inquiry was an appropriate decision, as is their clear-eyed critique of methods and clarity of argument in the field and in individual studies.

The CREDE synthesis work also exists in a context of domain interrelationships, so that many readers of this volume will find additional levels of resonance by reading the article-length reports of our other five synthesis teams (Systemic School Reform (Datnow, Lasky, Stringfield, & Teddlie, 2005); Families and Communities (Cooper, Chavira, & Mena, 2005); Preservice Teacher Education (Padron, 2005); Educating English Language Learners (Genesee, Lindholm-Leary, Saunders, & Christian, 2005); and Professional Development (Knight & Wiseman, 2005)), as well as the article-length version of the present volume (Lee, 2005).

A fine example of these domain interrelationships is Lee and Luykx's discussion of science education as an arena for the development of English language competence. The latter is the specific focus of the first volume in the Cambridge University Press series reporting CREDE's synthesis work

(Genesee, Lindholm-Leary, Saunders, & Christian, 2006).[3] Research indicates that the subject matter of science has rich potential as a setting for English language learning and that the techniques of sheltered instruction (reviewed in Genesee et al., 2006) offer illumination for teachers of science who wish to stimulate the learning of English. Similarly, the understanding of language learning and science instruction can inform those with a particular interest in systemic reform for schools with diverse student bodies.

CREDE's purpose was also to discern the underlying principles that can guide effective program design for diverse students. Built into CREDE's research design was the investigation of a set of principles extracted from previous research and development literature, which characterize successful educational programs for diversity. These principles were explored in all our research, to achieve a deep understanding of their dynamics and how they are expressed in diverse cultures. In our latest research program, these standards have been fully enacted at a programmatic level, and their effects measured against student achievement (e.g., Doherty, Hilberg, Epaloose, & Tharp, 2002; Doherty, Hilberg, Pinal, & Tharp, 2002; Doherty, Hilberg, & Tharp, 2003; Doherty & Pinal, 2004; Estrada, 2004).

We describe these principles as Standards for Effective Pedagogy (Tharp, Estrada, Dalton, & Yamauchi, 2000):

- I: Teachers and Students Producing Together (Joint Productive Activity). *Facilitate learning through joint productive activity among teacher and students.*
- II: Developing Language and Literacy Across the Curriculum. *Develop competence in the language(s) of instruction and of the disciplines throughout the day.*
- III: Making Meaning – Contextualizing School in Students' Lives. *Embed instruction in the interests, experiences, and skills of students' families and communities.*
- IV: Teaching Complex Thinking. *Challenge students toward cognitive complexity.*
- V: Teaching through Instructional Conversation. *Engage students through dialogue.*

Of course, these standards must be enacted within specific domains, content, and instructional goals. Readers familiar with the Effective Pedagogy Standards will find their understanding deepened by reading *Science Education and Student Diversity*, or indeed any of the other synthesis reports. In the learning of English, the learning of science, the learning to teach – there must finally be *content pedagogy*, in which the basic sociocultural human relationships of pedagogy are conditioned by the structures of knowledge

---

[3] Cambridge University Press will publish book-length versions of some of the other reports in this series.

present in each discipline. This interplay of levels of abstraction offers unparalleled intellectual stimulation and clear opportunities for further investigation of how we can draw ever closer to the goal of teaching all students.

In *Science Education and Student Diversity*, Lee and Luykx have held up a finely ground mirror, in which educators and researchers can see clearly our many achievements in learning how to bring young people of diverse backgrounds into an understanding and practice of science. Much of what we see here will make us proud. The authors serve us equally well by reminding us of what we still must discover, and how to do it.

Roland G. Tharp
Senior Scientist, Center for Research on Education,
Diversity & Excellence
Research Professor, University of California, Berkeley

References

Cooper, C., Chavira, G., & Mena, D. D. (2005). From pipelines to partnerships: A research synthesis on how diverse families, schools, and communities support children's pathways through school. *Journal of Education for Students Placed at Risk*, 10(4), 407–430.

Datnow, A., Lasky, S., Stringfield, S., & Teddlie, C. (2005). Systemic educational reform in racially and linguistically diverse contexts: A summary of the evidence. *Journal of Education for Students Placed at Risk*, 10(4), 441–453.

Doherty, R. W., Hilberg, R. S., Epaloose, G., & Tharp, R. G. (2002). Standards Performance Continuum: Development and validation of a measure of effective pedagogy. *Journal of Educational Research*, 96(2), 78–89.

Doherty, R. W., Hilberg, R. S., Pinal, A., & Tharp, R. G. (2002). *Transformed pedagogy, organization, and student achievement*. Paper presented at the annual meeting of the American Education Research Association, New Orleans, LA.

Doherty, R. W., Hilberg, R. S., & Tharp, R. G. (2003). Standards for effective pedagogy and student achievement. *NABE Journal of Research and Practice*, 96(2), 1–24.

Doherty, R. W., & Pinal, A. (2004). Joint productive activity, cognitive reading strategies, and achievement. *TESOL Quarterly*, 38.

Estrada, P. (2004). Patterns of language arts instruction activity: Excellence, fairness, inclusion, and harmony in first and fourth grade culturally and linguistically diverse classrooms. In H. C. Waxman, R. G. Tharp, & R. S. Hilberg (eds.), *Observational research in U.S. classrooms: New approaches for understanding cultural and linguistic diversity*. Cambridge: Cambridge University Press.

Genesee, F., Lindholm-Leary, K., Saunders, W., & Christian, D. (2006). *Educating English language learners*. Cambridge: Cambridge University Press.

Genesee, F., Lindholm-Leary, K., Saunders, W., & Christian, D. (2005). English language learners in U.S. Schools: An overview of research findings. *Journal of Education for Students Placed at Risk*, 10(4), 363–385.

Knight, S. L., & Wiseman, D. L. (2005). Professional development for teachers of diverse students: A summary of the research. *Journal of Education for Students Placed at Risk*, 10(4), 387–405.

Lee, O. (2005). Science education and student diversity: Synthesis and research agenda. *Journal of Education for Students Placed at Risk*, 10(4), 431–440.

Padron, Y. (2005). Final Report: Preservice education for teachers of diverse students: A research synthesis. Berkeley, CA: Center for Research on Education, Diversity & Excellence, University of California. www.crede.org

Tharp, R. G., Estrada, P., Dalton, S. S., & Yamauchi, L. A. (2000). *Teaching transformed: Achieving excellence, fairness, inclusion and harmony*. Boulder, CO: Westview Press.

Acknowledgments

We vividly remember the day we were contacted by Roland Tharp, Director of the Center for Research on Education, Diversity & Excellence (CREDE), and Thomas Carpenter, Director of the National Center for Improving Student Learning in Mathematics and Science (NCISLA). We were asked to analyze and synthesize the research on a range of issues regarding science education and student diversity, and then offer a research agenda. We accepted the task with both excitement and anxiety: excitement, from the thought that we would have an opportunity to learn more about this promising literature and to contribute to a knowledge base from which we hope will emerge more equitable learning environments for all students; and anxiety, from the realization that the book project would require a great amount of work; fair and balanced treatment of multiple, sometimes competing, perspectives in the literature; and realistic portrayals of both the accomplishments and the limitations of the colleagues, teachers, and students who are the subjects of our endeavors.

Just the process of doing a comprehensive review of literature was a daunting task. We especially appreciate the work of Margarette Mahotiere, who conducted the electronic search of the literature. Reading all the studies that met the criteria for inclusion in the book was equally daunting. Analyzing and synthesizing the literature on each of the relevant topics were formidable jobs. At times, writing about this vast body of research seemed impossible.

We were fortunate to have the support of the two national centers, CREDE and NCISLA in Mathematics and Science, and their respective directors, Roland Tharp and Thomas Carpenter. We were also fortunate to have the input of the Task Force members, who represented a range of academic disciplines and research areas:

Humberto Campins, University of Central Florida
Yolanda George, American Association for the Advancement of Science

Joseph Krajcik, University of Michigan
Julia Lara, Council of Chief State School Officers
Julio Lopez-Ferrao, National Science Foundation
Sharon Lynch, George Washington University
Sharon Nelson-Barber, WestEd
Trish Stoddart, University of California at Santa Cruz
Beth Warren, TERC

These valued colleagues helped at every step of the way, from the conceptualization of issues to feedback on draft versions of the manuscript. In particular, Sharon Lynch provided critical feedback on one of the later drafts of the book.

We are indebted to Cathy Murphy, editor at CREDE, who helped us with various draft versions of the manuscript. We also appreciate everyone at Cambridge University Press who worked on the book.

Our work was both strengthened and challenged by the fact that the two authors of the book complement each other in many ways. Okhee Lee is an educational researcher focusing on elementary science education and student diversity. Aurolyn Luykx is a linguistic and educational anthropologist with a critical theory orientation. Our differing perspectives would often lead us to argue over single words or phrases until we could come to an agreement. Our mutual respect for the differences in our academic training not only strengthened the manuscript but also deepened and broadened our respective understandings of educational research and practice. While collaboration across such differences inevitably implies many moments of frustration, these are far outweighed by the pleasures of colleagueship and intellectual growth that it provides.

Finally, we thank our families for their support and love.

Introduction

Ever since our nation first embraced the goal of mass schooling, it has faced the challenge of balancing the concern for educational quality with the desire to reach as many students as possible. Today, this dilemma is reflected in the dual aims of promoting high academic achievement while simultaneously pursuing educational equity for an increasingly diverse student population (Darling-Hammond, 1996; McLaughlin, Shepard, & O'Day, 1995). To achieve these aims, it is necessary to develop a knowledge base that situates recent advances in our understanding of educational processes within the realities of today's schools. This need is especially urgent, given the current climate of standards-based instruction, high-stakes assessment, and accountability. The literature review presented in this synthesis is a step in developing such an empirically based integration.

Knowledge about science and technology is increasingly important in today's world. Aside from the growing number of professions that require a working familiarity with scientific concepts and high-tech tools, the future of our society hangs in the balance of decisions that must be made on the basis of scientific knowledge. Documents on science education standards (American Association for the Advancement of Science [AAAS], 1989, 1993; National Research Council [NRC], 1996, 2000) represent the science education community's best efforts to define what constitutes science learning and achievement (see the summary in Lee & Paik, 2000; Raizen, 1998). According to these documents, science learning involves a two-part process: "to acquire both scientific knowledge of the world and scientific habits of mind at the same time" (AAAS, 1989, p. 190).

The development of scientific knowledge involves "knowing" science (i.e., scientific understanding), "doing" science (i.e., scientific inquiry), and "talking" science (i.e., scientific discourse). Knowing science involves making meaning of scientific concepts and vocabulary. One way that students come to know science is by doing science, that is, engaging

1

in science inquiry by generating questions, designing and carrying out investigations, analyzing data, proposing explanations, interpreting and verifying evidence, and constructing ideas to make sense of the world. Although knowing and doing have long been acknowledged as important components of science learning, recent science reform also emphasizes "talking science," whereby "teachers structure and facilitate ongoing formal and informal discussion based on a shared understanding of rules of scientific discourse. A fundamental aspect of a community of learners is communication" (NRC, 1996, p. 50).

The cultivation of scientific habits of mind entails adopting scientific values and attitudes, as well as the scientific worldview. Most cultural traditions embrace some values and attitudes that are associated with science, such as wonder, curiosity, interest, diligence, persistence, openness to new ideas, imagination, and respect toward nature. Other values and attitudes are particularly characteristic of Western modern science, for example, questioning, thinking critically and independently, reasoning from empirical evidence, making arguments based on logic rather than personal or institutional authority, openly critiquing the arguments of others, and tolerating ambiguity. Furthermore, science is a way of knowing that "distinguishes itself from other ways of knowing and from other bodies of knowledge" (NRC, 1996, p. 201). The scientific worldview is defined by a tradition of seeking to understand how the world works – to describe, explain, predict, and control natural phenomena. It is distinguished from alternative worldviews: "Explanations on how the natural world changes based on myths, personal beliefs, religious values, mystical inspiration, superstition, or authority may be personally useful and socially relevant, but they are not scientific" (NRC, 1996, p. 201).

Although the standards documents generally define science in the Western modern science tradition (AAAS, 1989, p. 136; NRC, 1996, pp. 201, 204), alternative views of science have been advocated by scholars in emerging areas of multicultural education, feminism, sociology and philosophy of science, and critical theory (Atwater & Riley, 1993; Calabrese Barton, 1998a; Eisenhart, Finkel, & Marion, 1996; Hodson, 1993; Lee, 1999a; Rodriguez, 1997; Stanley & Brickhouse, 1994, 2001). These scholars raise issues of power and the marginalization of nonmainstream groups, and challenge the very notion of science and the traditional definition of learning science (see the discussion in the section entitled "Views of Science: Is Science Independent of Culture?" in Chapter 2).

As immigrants, children of color, and children living in poverty come to represent an increasing fraction of the U.S. student population (García, 1999; National Center for Children in Poverty, 1995), science classrooms must address the educational needs of these children, who face the dual challenge of navigating the language and culture of the U.S. mainstream while also learning the academic norms, content, and processes of science disciplines. Thus, a vision of reform aiming at academic achievement for all students requires integrating disciplinary knowledge with knowledge of student diversity. Traditionally, disciplinary knowledge and student diversity have constituted separate research agendas. In the case of science education, although reform documents highlight "science for all" as the principle of equity and excellence (AAAS, 1989, 1993; NRC, 1996), they do not provide a coherent conception of equity or strategies for achieving it (Eisenhart et al., 1996; Lee, 1999a; S. Lynch, 2000; Rodriguez, 1997). On the other hand, the multicultural education literature emphasizes issues of cultural and linguistic diversity and equity, but with little consideration of the specific demands of different academic disciplines. In addition, although English language and literacy development in the context of subject area instruction is emphasized for English language learners – ELL students (Teachers of English to Speakers of Other Languages, 1997), research in this area focuses primarily on English language proficiency, with limited attention to achievement in subject areas such as science (August & Hakuta, 1997). Integration of "discipline-specific" and "diversity-oriented" approaches is necessary for achieving the goal of making science accessible for all students.

International studies, such as the Third International Mathematics and Science Study (TIMSS), reveal alarmingly poor performance of U.S. students on standardized science assessments (National Center for Education Statistics, 1996; Schmidt, McKnight, & Raizen, 1997). Additionally, the rank of U.S. students declines even further as they move up into the higher grades. Studies based on U.S. national samples, such as the National Assessment of Educational Progress (NAEP), indicate that the average scores for students of every age level and race/ethnicity have increased only slightly since the 1970s (Campbell, Hombo, & Mazzeo, 2000; O'Sullivan, Lauko, Grigg, Qian, Zhang, 2003; Rodriguez, 1998a). Furthermore, achievement gaps among students of diverse racial/ethnic and socioeconomic backgrounds have persisted in science achievement, as well as in science course enrollments leading to careers in science and engineering fields (Chipman & Thomas, 1987; National Science Foundation [NSF], 2002; Oakes, 1990).

Given overall poor science performance and the persistent gaps in science outcomes between mainstream and nonmainstream students in the United States, there is a pressing need to address students' cultural, linguistic, and socioeconomic circumstances in relation to science outcomes. Traditionally, while the science and science education communities advocate for greater participation of nonmainstream individuals in science-related fields, they expect these individuals to assimilate to the established institutional culture. There has been little recognition of the cultural and linguistic resources that nonmainstream individuals and groups bring to the science classroom, and little thought has been given to how to articulate these resources with the values and practices of science in order to enhance science outcomes in school and beyond.

Although classroom practices, local institutional conditions, and broader policy contexts affect all students, they are more likely to negatively impact nonmainstream students. All too often, teachers' knowledge of science and/or student diversity is insufficient to guide students from all backgrounds toward meaningful science learning. Furthermore, beginning teachers or those with inadequate teacher preparation tend to be assigned to inner-city schools where nonmainstream students are concentrated. Additionally, resources are scarcer and teacher attrition is higher in inner-city schools. Limited resources often force a trade-off between providing modified instruction that takes student diversity into account and reinforcing general standards to raise the quality of instruction for mainstream students (often to the detriment of other student groups). The trend toward standardization of curricula and assessment may also work against educational equity (McNeil, 2000), although there are efforts to promote both goals simultaneously (Delpit, 2003).

If we start from the assumption that high academic achievement is potentially attainable by most children, then achievement gaps among racial/ethnic, linguistic, or socioeconomic status (SES) groups can be interpreted as a product of (a) the learning opportunities available to different groups of students and (b) the degree to which circumstances permit them to take advantage of those opportunities. This poses questions for researchers and educators: What constitutes equitable learning opportunities, how do they vary for different student populations, and how can they be provided in a context of limited resources and conflicting educational priorities?

The literature reviewed in this book presents promising results about effective science education for nonmainstream students. These students come to school with already constructed knowledge, including their home language and cultural values. *Equitable learning opportunities* occur when school science values and respects the experiences these students bring from their home and community environments, articulates their cultural and linguistic knowledge with science disciplines, and offers educational resources and funding to support their learning at a level comparable to that available for mainstream students. Provided with equitable learning opportunities, these students are capable of demonstrating science achievement, interest, and agency, becoming bicultural and bilingual border crossers between their own cultural and speech communities and the science learning community.

This book analyzes and synthesizes current research on how cultural, linguistic, and socioeconomic factors in school and at home promote or hinder science achievement among nonmainstream K–12 students who have traditionally been underserved by the education system. Specifically, it examines how science achievement and other outcomes (broadly defined) are related to various factors involving science curriculum (including computer technology), instructional practices, assessment, teacher education, school organization, educational policies, and home and community connections to school science. The book emphasizes science education initiatives, interventions, or programs that have been successful with nonmainstream students. Based on the research synthesis, it proposes a research agenda to strengthen those areas in which the need for a knowledge base is most urgent, as well as those which show promise in establishing a robust knowledge base.

In analyzing and synthesizing current research, the book considers primarily peer-reviewed journal articles that provide clear statements of research questions, clear descriptions of research methods, convincing links between the evidence presented and the research questions, and valid conclusions based on the results (Shavelson & Towne, 2002). The rigor of the research methods employed is critically important in assessing the evidentiary warrants for the claims being made in each study and, more importantly, in assessing the robustness of a knowledge base in each area of research. The book provides descriptions of research methods along with results in each study, as well as discussion about methodological orientations and key findings in each area of research. The methodological and other criteria for the inclusion of research studies in the synthesis are described in detail in the Appendix.

There are four sections to the book, each with multiple chapters. In the first section, a range of conceptual and policy issues is addressed. The discussion starts with science achievement (i.e., measured outcomes) and student diversity as two key constructs in this synthesis. Based on this discussion, desired science outcomes for nonmainstream students are defined. Then, conceptual and policy issues guiding the synthesis are discussed, including the epistemological debate over definitions of science and school science, theoretical perspectives guiding research studies, and the policy context of high-stakes assessment and accountability in science education.

The second section starts with student characteristics and science learning linked to gaps in science outcomes among different student populations. Student learning occurs in the context of classroom practices – what materials are used, what content is taught, how the content is taught, and how students' mastery of the content is assessed. This section is organized into the following chapters: (a) student characteristics and science learning, (b) science curriculum (including computer technology), (c) science instruction, and (d) science assessment. Within each category, studies addressing bilingual or ELL students are discussed separately.

The third section addresses school- and home-based factors supporting or hindering science education in relation to gaps in science outcomes among different student populations. Classroom practices occur in the broader context of teacher education programs and educational policies. Although educational policies and practices influence all students, the impact is more consequential with nonmainstream students who are less likely to live in homes that provide the sort of academic supports that the school takes for granted. Thus, establishing connections between home/community and school science is critically important for nonmainstream students. This section consists of the following chapters: (a) science teacher education, (b) school organization and educational policy, and (c) home and community connections to school science. Within each category, studies addressing bilingual or ELL students are discussed separately.

Finally, we draw conclusions regarding two areas: (a) key features of the literature with regard to theoretical perspectives and methodological orientations, and (b) key findings about school- and home-based factors related to science outcomes of nonmainstream students. We offer recommendations for a research agenda to improve science outcomes and narrow achievement gaps among diverse student groups.

CONCEPTUAL GROUNDING AND POLICY CONTEXT

K nowledge of science and technology is an important part of being an educated citizen in the 21st century. As nonmainstream students come to constitute a large fraction of the nation's overall student population, achievement gaps in science among students of diverse cultural, linguistic, and socioeconomic backgrounds are of great concern. While achievement gaps in school science are generally comparable to those in other subject areas, science has not received as much attention from educators and researchers as have core subjects, such as reading, writing, and mathematics. Unlike literacy and numeracy, science is not perceived as a "basic skill"; this trend is reinforced by the fact that current policies of high-stakes assessment and accountability focus mainly on reading, writing, and mathematics. Furthermore, science is often ignored in inner-city schools (where nonmainstream students tend to be concentrated), due to limited funding and resources and the urgency of developing basic literacy and numeracy (National Center for Education Statistics, 1997).

1

Student Diversity and Science Outcomes

A focus on student diversity presumes that educational decisions, from statewide policies to individual classroom practices, may affect different student populations differently. Therefore, while the various aspects of student diversity are reflected in differing science outcomes, the ways in which policies and schools define, delimit, and manage student diversity may affect outcomes at least as much as does "diversity" itself. Regardless of the origin or nature of students' marginalization, academic success depends to a significant degree on assimilation to mainstream cultural and linguistic norms, for example, particular ways of structuring narratives, displaying competence, or interacting with adults, not to mention the phonological and grammatical conventions of standard English (Delpit, 1995; Heath, 1983). Traditional science instruction generally assumes that students have access to certain educational resources at home (such as computers, or adults with the time and academic skills to help with homework), and it requires students living in poverty to adopt learning habits that necessitate a certain level of socioeconomic stability (such as a quiet place to study, and freedom from child care or work-related responsibilities). While some students may overcome these barriers to academic success through exceptional talent, effort, or family support, the existence of such individuals does not negate the inequity of their educational circumstances or the need for social solutions to what are social, not individual, problems. Such issues must be taken into account in interpreting gaps in science outcomes among diverse student groups and in devising instructional programs to close the gaps.

Student Diversity

Student diversity in general, as well as particular categories of students, can be defined in different ways (Gutiérrez & Rogoff, 2003). This book focuses on student diversity in terms of race/ethnicity, culture, home language, and SES. This focus places particular emphasis on immigrant or U.S.-born racial/ethnic minority students, whose educational success depends largely on acquiring the standard language and shared culture of "mainstream" U.S. society. Most of these students are characterized as non-White, and a disproportionate number come from low-income families (August & Hakuta, 1997; García, 1999; National Center for Children in Poverty, 1995).

While categories such as these are necessary for analytic reasons, they are heuristic tools rather than natural groupings or fixed human characteristics. Researchers aiming to shed light upon the social patterning of educational access and achievement, as well as readers of educational research, should keep in mind that the social reality in which educational processes occur is inevitably more complex than such categorical divisions imply. For this reason, the following caveat by Lemke (2001) is pertinent, though infrequently observed:

> I should not be using terms such as *class, gender, sexuality*, and especially *race*, or even in many contexts *culture* and *language*, without problematizing them. None of these notions has objective definitions; all of them represent potentially misleading and harmful oversimplications of the complexity of human similarities and differences. All of them owe their origins and historical prominence to explicitly political rather than scientific agendas. Every research study which frames itself in these terms should also be an inquiry into the limitations of applicability of the concepts themselves, refining and replacing them according to the salient features of the data at hand. Every researcher who uses them should have investigated their histories and be familiar with the relevant critiques of their validity. This is not often enough the case in the science education literature. (p. 303)

Each of the dimensions of identity named here – race/ethnicity, culture, language, and social class – is itself a complex, shifting, social, and political field. At the same time, the interplay among them is also complex. On the one hand, it is difficult methodologically to separate out the influences of different variables, which may cut across populations in ways that are not easily untangled. For example, a given immigrant population may contain individuals of varied racial/ethnic backgrounds, racial/ethnic groups are internally stratified by class, and certain cultural values and practices may be shared across different socioeconomic strata within a racial/ethnic group while others may not (e.g., Lee, 1999b). On the other hand, these variables are not entirely independent of one another, conceptually speaking; language is an important element of race/ethnicity, culture is partly determined by social class, and so on. Racial/ethnic identities as well as language proficiencies are less discrete than is implied by commonly used demographic categories; they may vary within a single household or across the life-span of a single individual. Furthermore, although shared language, culture, and ancestry are generally important components of racial/ethnic identity, the relative importance of each component varies widely from one racial/ethnic group to another and from one social context to another.

Social theorists have proposed concepts such as "languaculture" (Agar, 1996), "class cultures" (Bourdieu, 1984), "social class dialects" (Labov, 1966), and even "Ebonics" (Ogbu, 1999) to capture the inevitable intertwining of race/ethnicity, culture, language, and social class – not to mention the complex ways in which gender interacts with all of these areas.[1] Especially with regard to native speakers of nonstandard dialects of English (e.g., African Americans, working-class Whites, and some Hispanic and Native American populations), the influences of race/ethnicity, culture, language, and social class on students' educational performance are more often conflated than systematically analyzed. Failure to disaggregate student outcomes according to these variables has limited the knowledge base with regard to the educational progress of nonmainstream students. On the other hand, the habit of treating these variables as discrete and independent, both conceptually and methodologically, and the failure of most educational research to adequately theorize the connections among them, has further limited research.

Varying usages of terminology to refer to human social groups often reflect different theoretical stances or disciplinary traditions. This is particularly notable with regard to racial/ethnic categories. Most social scientists today agree that human "races" are cultural categories rather than biological ones (American Anthropological Association, 1998). This is evident from the fact that racial groupings are defined differently from one society to another. Nevertheless, governmental bureaucracies, including educational systems, continue to treat them as discrete, self-evident designations, with the result that children may be categorized differently from their parents, or children of different nationalities may be lumped together in the same statistical category on the basis of their skin color.

The lack of consensus around demographic designations for different categories of students reflects the rapidly changing makeup of the population, the changing political connotations of different terms, and the specific aspects of identity that researchers and/or subjects may wish to emphasize. Although this sometimes causes difficulty with regard to comparability of studies, the lack of a standard terminology to describe the overlapping dimensions of student diversity is a valid reflection of the fluid, multiply determined, and historically situated nature of identity, and the ways in which such designations are used to stake out particular claims about the location and nature of social boundaries. While much of the science

---

[1] The science education literature on gender as it intersects with race/ethnicity, culture, language, and social class is limited and is not discussed in this report (see Baker & Leary, 1995; Brickhouse & Potter, 2002; Brickhouse, Lowery, & Schultz, 2000; Catsambis, 1995; Davis, 2002; Jegede & Okebukola, 1992; Rennie, 1998).

education literature (especially those studies based on quantitative analysis of student outcome data) tends to treat such categories as unproblematic, this should be understood as a necessary fiction that makes possible the management of large data sets to reveal "the big picture" with regard to student diversity and science outcomes. In reality, the number of students whose personal circumstances cross and confound such categorical boundaries is greater than ever, and will no doubt continue to increase as those boundaries become more flexible and porous.

Throughout this book, the terms "mainstream" and "nonmainstream" are used with reference to students' racial/ethnic, cultural, linguistic, and socioeconomic backgrounds. Similar to contemporary usage of the term "minority" by social scientists, "mainstream" is understood to refer not to numerical majority, but rather to social prestige, institutionalized privilege, and normative power. Thus, in classroom settings, "mainstream" students (i.e., those who are White, middle or upper class, and native speakers of standard English) are more likely than "nonmainstream" students to encounter ways of talking, thinking, and interacting that are continuous with the skills and expectations they bring from home, and this continuity between home and school constitutes an academic advantage relative to nonmainstream students. These group-level phenomena may not apply to particular individuals or may be offset by other factors, such as proficiency levels in both the home language and English, immigration history, degree of acculturation, parents' educational levels, and family/community attitudes toward education in general and science education in particular. Recognizing overall differences between groups does not justify limiting one's expectations of individual students, but it does provide a framework for interpreting observed patterns and processes that occur with differing frequency among different groups (Gutiérrez & Rogoff, 2003).

The more inclusive terms "diverse student groups" and "students from diverse backgrounds" are used to refer to the entire gamut of students, mainstream and nonmainstream. This usage is intended to emphasize several key points. First, the frequent use of these terms to refer exclusively to nonmainstream students sets those students apart as deviating from an assumed norm in supposedly similar ways, while defining middle-class White students as a normative, collectively homogeneous ideal; such suppositions are increasingly untenable within contemporary social science. Second, designating only nonmainstream students as "culturally and linguistically diverse" implies that only those students bring cultural and linguistic "baggage" to the classroom, whereas in reality, culture and language play as large a role in the educational experience of mainstream students as in that of nonmainstream students. Third, the "mainstream" can no longer be assumed to be representative of most students' experience, especially in inner-city schools or large urban school districts where nonmainstream students make up the majority. For these reasons, we use "student diversity" to refer to the full range of variability, rather than just the stigmatized elements of the student population.

The terms "first language," "home language," "native language," and "mother tongue" are used interchangeably in this book, since their use among the studies reviewed herein is not consistent. While the language that many immigrant families use predominantly at home may not actually be their first language, none of the studies considered in this book took such nuances into account. Similarly, the variability in the terms referring to differing racial/ethnic categories reflects the variable usage of different researchers.

Terminology can be problematic in any synthesis because some researchers use established terms to mean different things, while others invent their own terms to express novel concepts (or rejection of existing terms). In this book, terms are used as they appeared in the studies in order to represent the original intentions of the researchers, to the extent that this does not confuse or conflate the ways these terms are typically used in the literature.

Gaps in Science Outcomes

Science outcomes are defined broadly to include science achievement, attitudes toward science, enrollments in high school science courses, earning of college and graduate degrees in science and engineering fields, and entrance into science and engineering occupations. The results of national and international science assessments, described in this section, indicate that closing the gaps in science outcomes among racial/ethnic and socioeconomic groups, as well as improving the science outcomes of all students, must constitute the dual goals of science education in our nation. Thus, there is a critical need for a knowledge base sufficient to explain these gaps and to design educational programs to close them.

Ideological and Methodological Limitations

Descriptions of science achievement gaps must be interpreted within the context of ideological and methodological limitations in the current knowledge base. In the ideological sense, A. Rodriguez (1998a) argues that failure to disaggregate science achievement data – for example, by socioeconomic strata within racial/ethnic groups or by subgroups within broad racial/ethnic categories – may create or reinforce stereotypes about a certain group. For instance, the "model minority" stereotype for Asian American students, particularly as regards their performance in mathematics and science, masks great disparities and challenges facing many students, such as Southeast Asian refugees with little schooling or limited literacy development in their home countries. In contrast, high-achieving Hispanic students may be at a disadvantage because teachers and other school personnel tend to have low expectations of their academic ability.

Rodriguez (1998a) addresses achievement gaps among racial/ethnic groups in terms of social justice in the education system at large. Contrary to the notion of meritocracy – whereby all students who work hard get proper rewards – Rodriguez's analysis of the achievement data suggests that the education system is structured so as to benefit those groups already in power, inasmuch as it rewards prior membership in powerful social groups, rather than some objective ideal of "merit." In contrast, the students most adversely affected by the meritocracy myth come from the fastest growing racial/ethnic groups. Rodriguez claims that to promote participation and achievement of nonmainstream students in science, the meritocracy myth needs to be exposed and dealt with.

Other scholars as well have considered achievement gaps within the context of larger power structures. According to G. Madaus (1994), equitable assessment suggests the ideal of "just" measures in educational and social conditions. He argues that although assessment technology is not by nature socially unjust, it is intertwined with the distribution of wealth, racial/ethnic hierarchies, and gender relations. As more advanced technologies in assessment are introduced and developed, elite groups determine what is to be assessed and how. Elites control not only the gatekeeping process of assessment but also the decision-making process around assessment criteria and practices, and they tightly restrict public access to information about this process. Thus, new technologies create new social problems, despite their potential benefits. The negative impacts of these technologies are most deleterious to those groups that are already marginalized. Madaus cautions that "testing as a technology has the potential to perpetuate current social and educational inequalities" (p. 76).

As for methodological limitations, achievement is typically measured by standardized tests administered to national and international student samples. These databases provide overall achievement results by race/ethnicity, SES, and gender, but contain very limited information with regard to disaggregation of results, such as socioeconomic strata within racial/ethnic groups or subgroups within broad racial/ethnic categories. For example, Mexican Americans, Chicanos, Puerto Ricans, and students of various other nationalities are often collapsed together under the generic category of "Hispanics." This lack of information impedes researchers and policymakers from gaining an accurate understanding of achievement gaps among specific student groups.

ELL students were excluded from most large-scale assessments until very recently. While the 2000 National Assessment of Educational Progress report card was the first (since the NAEP's inception in 1969) to analyze assessment accommodations in science, the results did not disaggregate limited English–proficient students from students with disabilities (O'Sullivan et al., 2003). This practice "literally creates a kind of systemic 'ignorance' about [ELL students'] educational progress" and "leaves the school, district, or system utterly unable to account for the learning of these students" (Lacelle-Peterson & Rivera, 1994, p. 70).

Science Achievement

U.S. students do not rank favorably on international measures of science achievement (for a summary, see S. Lynch, 2000; Rodriguez, 1998a). In the largest study of its kind, the 1995 TIMSS found that U.S. students were far from the top in science relative to their counterparts in other countries (National Center for Education Statistics, 1996; Schmidt et al., 1997). While U.S. 4th-grade students scored within the cluster of top-performing nations, 8th-grade students scored only slightly above the international average, and 12th-grade students scored among the lowest performing nations. In the 1999 TIMSS-Repeat (TIMSS-R), which involved 8th-grade students only, U.S. students ranked slightly above the international average (Martin et al., 2001). Of the 14 U.S. school districts participating in the TIMSS-R, all 4 urban school districts performed significantly below the international average.

At the national level, the long-term trend assessments of U.S. students in science, as measured by the NAEP, indicate that the average scores for students of every age level and race/ethnicity have increased slightly since the 1970s (Campbell et al., 2000). Between 1977 and 1999, the average score for 9-year-old students rose from 220 to 229; for 13-year-old students, from 247 to 256; and for 17-year-old students, from 290 to 295 (see Table 1). The scale ranges from 0 to 500.

Achievement gaps among racial/ethnic groups as indicated by the NAEP are gradually narrowing, as scores of Black and Hispanic students have improved since the 1970s at a slightly faster rate than have the scores of White non-Hispanic students. Nevertheless, Black and Hispanic students' scores remain well below those of White students at each grade level. Additionally, achievement growth rates of students by race/ethnicity and gender indicate that with the exception of Hispanic males, the growth rates of African American and female Hispanic students are so minimal that their final 12th-grade achievement level still fell well below the initial 8th-grade achievement of Whites and Asian Americans (Muller, Stage, & Kinzie, 2001). In other words, White and Asian American 8th-grade students' science achievement was generally similar to that of 12th-grade African American and female Hispanic students.

While the long-term trend assessments of science achievement by SES are not available (since scores are categorized only by students'

TABLE 1. *Average Science Proficiency by Age and Race/Ethnicity of Students: 1977 to 1999*

Race/Ethnicity	1977	1982	1986	1990	1992	1994	1996	1999
9-Year-Olds								
Total	220	221	224	229	231	231	230	229
White, non-Hispanic	230	229	232	238	239	240	239	240
Black, non-Hispanic	175	187	196	196	200	201	202	199
Hispanic	192	189	199	206	205	201	207	206
13-Year-Olds								
Total	247	250	251	255	258	257	256	256
White, non-Hispanic	256	257	259	264	267	267	266	266
Black, non-Hispanic	208	217	222	226	224	224	226	227
Hispanic	213	226	226	232	238	232	232	227
17-Year-Olds								
Total	290	283	289	290	294	294	296	295
White, non-Hispanic	298	293	298	301	304	306	307	306
Black, non-Hispanic	240	235	253	253	256	257	260	254
Hispanic	262	249	259	262	270	262	269	276

race/ethnicity or gender), the 1996 and 2000 NAEP results indicate that students who were eligible for the free/reduced-price lunch program performed well below those who were not eligible (O'Sullivan et al., 2003). On the composite scale ranging from 0 to 300, the differences ranged from 23 points for grade 8 in 1996 to 29 points for grade 4 in 2000. From 1996 to 2000, the achievement gaps widened by three points for grade 4 and nine points for grade 8, while the gap narrowed by five points for grade 12. Again, it is difficult to pinpoint the factors accounting for these gaps, since students of color are disproportionately represented in free/reduced-price lunch programs and the interaction between variables such as race/ethnicity and SES was not analyzed.

Rodriguez (1998a) conducted a systemic analysis of trends in science achievement by race/ethnicity, SES, and gender, using national databases including the NAEP, National Education Longitudinal Study (NELS), American College Test (ACT), Scholastic Aptitude Test (SAT), and Advanced Placement (AP) Exams. The results indicated improvement for all student groups in science achievement and participation, but wide gaps persisted between Anglo-European students and students from African and Latino groups (to use Rodriguez's terms). In addition, patterns of achievement gaps were alarmingly congruent over time and across studies with respect to race/ethnicity, SES, gender, and grade level.

Science Attitudes

Attitudes toward science vary among racial/ethnic groups. Furthermore, this variation is not always consistent with the variation in science achievement. Studies from the United States have reported positive attitudes toward science among nonmainstream students. S. J. Rakow (1985a) used a modified version of the 1981–1982 NAEP with a stratified sample of about 2,000 9-year-olds, about 7,900 13-year-olds, and about 8,000 17-year-olds, who were randomly selected from across the nation. The study involved six subgroups based on race/ethnicity and gender. Results about science content, inquiry skill, and attitudes for each subgroup were reported in comparison to the national average. At each age, White males and females had higher scores for science content and inquiry skill than did Black and Hispanic males and females. In contrast, there were no noticeable patterns of difference in science attitudes among the racial/ethnic and gender subgroups.

J. B. Kahle (1982) used the 1976–1977 NAEP items concerning attitudes toward science with 13- and 17-year-olds. (The sample size for each age group is not provided in the article.) The results indicate that although Black students had lower science achievement on the NAEP than did their White counterparts, they expressed more positive attitudes toward science, science classes, and science careers. However, they had fewer science-related experiences, found science less useful outside of school, and were less aware of scientific methods and of how scientists work than their White counterparts.

M. M. Atwater, J. Wiggins, and C. M. Gardner (1995) examined more than 1,400 urban middle school students' attitudes toward science and science-related careers. The study involved predominantly African American students in grades 6, 7, and 8 at 3 middle schools, from a range of SES backgrounds. The 3 schools constituted a stratified random sample selected from the 10 middle schools in a large urban school district in the southeastern part of the United States. The study applied four subscales based on an existing attitude instrument: general self-concept, achievement motivation in school, science self-concept, and science anxiety. The results indicate that most students had very high general self-concept and high achievement motivation in school. In contrast, their science self-concept was not as high, and most students were somewhat anxious about science. Additionally, a majority of the students expressed an uncertain attitude toward science curricula and science teachers. Less than 50% of the students indicated any interest in engaging in science at the high school level, although many planned to enter science-related careers.

The results of Rakow (1985a) and Kahle (1982) indicate that nonmainstream students have positive attitudes toward science and aspire to enter careers in science. However, compared to their White counterparts, they have lower science achievement and inquiry skills, as well as limited exposure and access to the knowledge necessary to realize such aspirations. Furthermore, M. M. Atwater et al. (1995) report that a majority of African American middle school students expressed an uncertain attitude toward science curricula and did not indicate an interest in engaging in high school science. The results raise important questions concerning the reasons underlying discrepancies between nonmainstream students' attitudes and their actual achievement in science, ways that educators might tap into students' positive attitudes to improve science achievement, and what happens to these positive attitudes if students do not eventually experience academic success in science.

High School Science Course Enrollment, College Major, and Career Choice

Other indicators of science outcomes include science course enrollments, college major, and career choice. While this book is limited to research on K–12 education, the decision to major in a scientific field in college or to pursue a career in science or engineering is largely influenced by science learning experiences throughout elementary and secondary school. Overall, racial/ethnic minority groups have made substantial gains in science and engineering fields from high school and beyond, but gaps persist (Chipman & Thomas, 1987; NSF, 2002; Oakes, 1990). In order to place the following demographic data in perspective, it is useful to remember that among U.S. citizens and permanent residents, Whites account for 72%, Blacks 12%, Hispanics 12%, Asian Americans (including Pacific Islanders) 4%, and American Indians (including Alaskan Natives) less than 1% (NSF, 2002, p. 2).

With regard to enrollments in high school science courses, in both 1990 and 1998, most students across racial/ethnic groups had taken biology, whereas relatively few had taken AP/honors biology or engineering (NSF, 2002, pp. 6, 104; see the summary in Table 2). Racial/ethnic differences are noticeable with regard to enrollments in high school chemistry and physics. In 1990, smaller percentages of Black, Hispanic, and American Indian high school graduates had taken chemistry and physics, compared to White and Asian American graduates. By 1998, the percentages for Black, Hispanic, and American Indian graduates had increased, but the percentages of White and Asian American graduates had also increased comparably.

Throughout the 1990s, racial/ethnic minority groups made gains with regard to both absolute numbers and percentages of bachelor's, master's, and doctoral degrees awarded in science and engineering fields (NSF, 2002; see pp. 31–33, 153–154 for bachelor's degrees; pp. 47–48, 202–204 for master's degrees; and pp. 51–54, 217–222 for doctoral degrees). However,

TABLE 2. *Percentages of High School Graduates Taking Selected Science Courses in High School, by Race/Ethnicity*

	1990					1998				
Course	White	Black	Hispanic	Asian/ Pacific Islander	American Indian/ Alaskan Native	White	Black	Hispanic	Asian/ Pacific Islander	American Indian/ Alaskan Native
Biology	91.3	91.1	90.1	90.4	89.4	93.7	92.8	86.5	92.8	91.3
AP/honors biology	10.5	7.7	6.7	13.4	3.8	16.7	15.4	12.6	22.2	6.0
Chemistry	51.4	40.0	38.1	63.6	34.9	63.2	54.3	46.1	72.4	46.9
Physics	23.1	14.5	13.2	38.4	14.5	30.7	21.4	18.9	46.4	16.2
Engineering	4.8	1.4	0.8	3.4	3.2	7.9	4.8	2.3	5.2	9.6

TABLE 3. *Percentages of Degrees Awarded to U.S. Citizens and Permanent Residents in Science and Engineering, by Race/Ethnicity*

Degree	1990						1998					
	White	Black	Hispanic	Asian/Pacific Islander	American Indian/Alaskan Native	Unknown	White	Black	Hispanic	Asian/Pacific Islander	American Indian/Alaskan Native	Unknown
Bachelor's	81.0	5.5	4.2	5.8	0.4	3.1	73.2	7.9	6.7	8.7	0.6	2.8
Master's	79.5	3.3	2.8	7.3	0.3	6.7	74.6	5.7	4.7	9.4	0.5	5.1
Doctorate	85.7	2.4	3.0	6.6	0.3	2.0	78.4	4.1	3.9	11.2	0.7	1.7

TABLE 4. *Percentages of Employed Scientists and Engineers, by Race/Ethnicity*

1993					1999				
White	Black	Hispanic	Asian/Pacific Islander	American Indian/Alaskan Native	White	Black	Hispanic	Asian/Pacific Islander	American Indian/Alaskan Native
84.1	3.6	2.9	9.1	0.2	81.8	3.4	3.4	11.0	0.3

gaps among racial/ethnic groups persist. The information about relative percentages, including only U.S. citizens and permanent residents, in 1990 and 1998 is summarized in Table 3.

Not surprisingly, Whites predominate in science and engineering occupations (NSF, 2002, pp. 61, 249–250; see the summary in Table 4). Asians/Pacific Islanders are overrepresented in science occupations, whereas Blacks, Hispanics, and Native Americans are underrepresented. This pattern (including only U.S. citizens and permanent residents) did not change much between 1993 and 1999.

Y. S. George et al. (2001) reviewed more than 150 research efforts related to choice of and retention in college majors; academic mentoring at both precollege and higher education levels; pursuit of a doctorate; and faculty positions in science, technology, engineering, and mathematics (STEM). They examined the reasons why the representation of minorities and women in STEM lags far behind that of White men, despite programs promoting educational opportunities over the past 25 years. They discussed key research efforts, identified gaps, and proposed a research agenda, with particular attention to the transitions from one level of academic achievement to the next. They identified three priorities in education or social science research in order to promote participation of minorities and women underrepresented in STEM from the high school years to the professoriate: (a) improve research methodologies, (b) foster linkages between different types of research, and (c) explore new research areas.

Science Outcomes

As mentioned earlier, science outcomes are defined in broad terms that include achievement scores on standardized tests, course enrollments, high school completion, higher education, and career choices in science and engineering fields. Science outcomes also include meaningful learning of classroom tasks, as well as affect (attitudes, interest, motivation) in science.

In the current policy context, which stresses structured English immersion for ELL students (without attention to the development of the student's first language) and severely limits content area instruction in languages other than English, English proficiency becomes a de facto prerequisite for science learning. In this sense, acquisition of oral and written English and exit from English as a Second Language (ESL) or English to Speakers of Other Languages (ESOL) programs, while they do not constitute "science outcomes" per se, play a large role in determining science outcomes as they are commonly measured.

For students from nonmainstream backgrounds, desired science outcomes include becoming bicultural, bilingual, and biliterate with regard to the home language and culture, on the one hand, and Western science, on the other. Students from all language backgrounds need to acquire the discourse of science as well as the discourse of their homes and communities in order to understand the culture of science as well as their own culture, and to behave competently across social contexts. Furthermore, from a critical theory perspective, desired science outcomes include social activism, as nonmainstream students become aware of social injustice and inequity – the unequal distribution of social resources and the school's role in the reproduction of social hierarchy – and take actions to address this problem in their communities.

As discussed throughout the book, current educational policies and practices do not generally support these outcomes. By focusing on assimilating nonmainstream students to the mainstream, policies and practices do not substantially engage or incorporate the knowledge, traditions, and practices of nonmainstream groups. By focusing on students' acquisition of English and neglecting the maintenance and/or development of students' oral and written proficiencies in the home language, policies and practices fail to take advantage of the intellectual resources that ELL students bring to the classroom. Furthermore, by indoctrinating nonmainstream students into existing power structures, policies and practices do not allow them to develop a critique of the social injustice and inequity in which they live.

While debates over the most useful ways of conceiving student diversity, science outcomes, and science itself remain unresolved, participants in these debates generally agree that nonmainstream students have not been effectively served by school science as it has traditionally been taught, and that the reasons for this are socially patterned rather than being explicable in terms of individual variables. Starting from this premise, researchers have sought to examine the nature of science and science instruction, as well as the various dimensions of student diversity. The following chapters in this book provide a conceptual and empirical map to the state of science education research as it relates to student diversity, in the areas of learning, curriculum, instruction, assessment, teacher education, school organization, educational policies, and school connections to students' home and community life. Each of these areas raises specific questions and challenges with regard to the goal of providing equitable learning opportunities in science for all students.

2

Conceptual Frameworks and Educational Policies

This research synthesis is framed by conceptual considerations, discussed in this chapter, as well as by methodological approaches (see Appendix). Three conceptual and policy issues that guide the work reviewed here are: (a) perspectives on the epistemology of science, that is, what counts as science and school science; (b) theoretical perspectives guiding the research literature; and (c) the policy context of high-stakes assessment and accountability.

Views of Science: Is Science Independent of Culture?

One strand of the debate over science education among diverse student groups has focused on epistemological questions, such as "What counts as science?" and "What are scientific ways of knowing?" The definition of science is "a de facto 'gatekeeping' device for determining what can be included in a school science curriculum and what cannot" (Snively & Corsiglia, 2001, p. 6; also see Cobern & Loving, 2001; Loving, 1997; Siegel, 2002; Stanley & Brickhouse, 1994, 2001).

The core of this epistemological debate involves universal versus multicultural views of science (Brickhouse, 1994; Cobern & Loving, 2001; Loving, 1997; Siegel, 1995, 2002; Snively & Corsiglia, 2001; Southerland, 2000; Stanley & Brickhouse, 1994). "Science" has traditionally been equated with Western science over the last 500 years (AAAS, 1989, p. 136, and 1993; NRC, 1996). This definition conceives of science in universal terms – "Science assumes that the universe is, as its name implies, a vast single system in which the basic rules are everywhere the same. Knowledge gained from studying one part of the universe is applicable to other parts" (AAAS, 1989, pp. 3–4). Universalism considers Western modern science, while originating in a specific cultural tradition, to be a universally valid endeavor with a set of tenets that transcends cultural boundaries.

Multicultural science educators criticize the traditional assumption that science is universal and "culture-free" and claim that such a view fails to consider other cultures' views of the natural world (Atwater, 1996; Eisenhart et al., 1996; Lee, 1999a; Rodriguez, 1997). The multicultural science literature conceives of science as a socially and culturally constructed discipline, questions the dominance of Western modern science, and advocates for inclusion of non-Western, indigenous, or other racial/ethnic traditions of knowing the natural world. They base their claims on several extensive bodies of literature.

First, multiculturalists emphasize ethnoscience, or notions of "how the local world works through a particular cultural perspective" (Snively & Corsiglia, 2001, p. 10). They highlight a rich and well-documented indigenous knowledge base, known to biologists, ecologists, and anthropologists as "traditional ecological knowledge," that has sustained indigenous populations over many centuries by providing pragmatic local practices organized around the relationship between environmental processes and human needs (McKinley, 2004; Riggs, 2005; Snively & Corsiglia, 2001).

Second, they point out that ethnoscientific traditions have produced knowledge relevant to the Western scientific and technological disciplines of agriculture, botany, ecology, medicine, astronomy, navigation, climatology, and architecture, among others. Furthermore, they argue for including the contributions of non-Western cultures in the history of Western modern science (e.g., Hodson, 1993; Krugly-Smolska, 1996; Murfin, 1994). They claim that recognizing the scientific and technological contributions of other cultures provides a broader view of what science is and what it represents, and fosters the advancement of Western modern science.

Third, on the basis of a large body of literature concerning worldviews across various cultures within the United States and around the world, they use the plural "sciences" to refer to multiple ways of understanding the natural world (e.g., Kawasaki, 1996; Ogawa, 1995; Stanley & Brickhouse, 1994, 2001). Based on W. W. Cobern's (1991) comprehensive framework, scholars have identified alternative worldviews associated with diverse languages and cultures, which are sometimes incompatible with the scientific worldview. (The literature on worldviews is described in detail in the next chapter.)[1]

---

[1] Due to the focus on nonmainstream students, this synthesis does not include studies on how the worldviews of mainstream students may influence their science learning. Though Western science is more closely associated with the U.S. mainstream than with nonmainstream communities, recent years have seen an intensification of the epistemological and political tensions between the scientific community and certain religious sects that are increasingly considered "mainstream" in our nation. The growing political dominance of such sects has already had a notable effect on science education, particularly with regard to topics such as evolution and sexuality. While it is beyond the scope of this book to address

Fourth, they justify multicultural science based on the principle of moral justice (Irzik & Irzik, 2002; Siegel, 1997). They argue that multiculturalism requires a universal or transcultural commitment to treat members of all cultures justly and with respect. From this perspective, cultural domination, hegemony, and nonrecognition of others is morally wrong for all cultures, and multiculturalism should be embraced by all, even by those who do not yet recognize this as a moral truth. They further argue that multiculturalism is compatible with epistemic universalism, and the two can peacefully coexist at the moral level.

Finally, they justify multicultural science based on antiracism (Carter, 2004; Hodson, 1999; Hodson & Dennick, 1994). Antiracism is concerned with revealing and combating racist attitudes and practices that disadvantage and discriminate against people of color, resulting in an unequal distribution of opportunity, wealth, and power. This position claims that the multiculturalist and antiracist approaches are reconcilable if emphasis is placed on both celebration of diversity and awareness of racism. It further claims that the history of science and technology has a special role in the antiracist approach that is an essential element of equitable science education for a multiethnic and multicultural society.

Although the multicultural science literature values both Western modern science and alternative views of the natural world, relative emphasis differs along the spectrum of radical to moderate approaches. Radical approaches based on critical theory argue that the nature and practice of science, as traditionally defined by middle-class white males, should be transformed to include multiple voices and ways of knowing characteristic of female and non-Western participants (Calabrese Barton, 1998a, 1998b; Rodriguez, 1997). On the other hand, moderate approaches recognize and aim to integrate the beliefs and worldviews of non-Western peoples, while emphasizing the explanatory and predictive power of Western modern science and ways of knowing (Aikenhead & Jegede, 1999; Lee, 1999a; Lee & Fradd, 1998; Loving, 1997).

Despite such variations, the multicultural science literature expresses the concern that universalism grounded on Western modern science may lead to assimilation, as it expects students to identify with science as universal knowledge and to leave their cultural beliefs behind in order to succeed in the dominant society. W. B. Stanley and N. Brickhouse (1994) argue that "science education has remained immune to the multiculturalist critique by appealing to a universalist epistemology; that the culture,

---

this controversy in any detail, we would be remiss if we failed to note the degree to which antiscience and anti-intellectual views already shape our cultural and political life, as well as our nation's science policy. In short, the changing "mainstream" worldview does not always coincide with the scientific worldview, and scientists themselves, though mostly products of the "dominant society," often find themselves at odds with that society.

gender, race, ethnicity, or sexual orientation of the knower is irrelevant to scientific knowledge" (p. 388).

Even within the tradition of Western modern science, different academic disciplines have different views of scientific practices, ranging from the sociology of science's focus on political and human endeavor (Cunningham & Helms, 1998; Kelly, Carlsen, & Cunningham, 1993; Kuhn, [1970] 1996) to more mechanistic schools of thought that deemphasize the influence of such factors. Some critics, referring to recent research in the sociology of science, argue that the views of science and scientific practices put forth in standards documents (AAAS, 1989, 1993; NRC, 1996) do not reflect the practices of actual scientific communities as portrayed by B. Latour and S. Woolgar (1986), M. Lynch (1985), and E. Ochs, S. Jacoby, and P. Gonzales (1996). They contend that standards documents present a narrow, distorted, or truncated view of scientific practices, and argue instead for a view of science as flexible, fluid, interactive, and reflexive (Brown, 1992; Lehrer & Schauble, 2000; Metz, 1998; Warren et al., 2001).

While recognizing the existence of multiple views of science, this book focuses on school science as implicitly defined in standards documents, that is, as the systematic search for empirical explanations of natural phenomena (AAAS, 1989, 1993; NRC, 1996; see the summary in Lee & Paik, 2000; Raizen, 1998). The goal of school science is for students to develop an understanding of key science concepts, engage in scientific inquiry and reasoning, participate in scientific discourse, and cultivate scientific habits of mind such as scientific values and attitudes as well as the scientific worldview (see the description in the Introduction).

Theoretical Perspectives Guiding This Synthesis

This book emphasizes the view that learning is mediated by cultural, linguistic, and social factors. Learning is enhanced – indeed, made possible – when it occurs in contexts that are culturally, linguistically, and cognitively meaningful and relevant to students. If their home languages and cultures are not considered in the educational process, schooling ignores or even negates the tools that students have used to construct their understandings of the world. It is these prior understandings that provide a meaningful context for the construction of new understandings. Thus, effective science instruction incorporates students' prior cultural and linguistic knowledge in relation to science disciplines.

The research studies reviewed throughout this book address the intersection of science disciplines with students' race/ethnicity, culture, language, and social class from a range of theoretical perspectives and methodological orientations. Science educators have proposed various theoretical underpinnings to guide research and practice, such as a multicultural perspective on the epistemology of science (Brickhouse, 1994;

Snively & Corsiglia, 2001), a cross-cultural perspective on worldviews (Aikenhead & Jegede, 1999; Cobern, 1991, 1996), a social constructivist perspective on multicultural science education (Atwater, 1993, 1996), sociocultural perspectives more generally (Lemke, 2001; O'Loughlin, 1992), postmodern perspectives (Haraway, 1990, 1991; Norman, 1998), critical theory (Calabrese Barton, 1998a, 1998b; Rodriguez, 1998b; Seiler, 2001; Tobin, Seiler, & Smith, 1999), and a civil rights and social justice perspective (Tate, 2001).[2] Rather than interpreting issues of science teaching and learning from a particular theoretical perspective, this book considers research originating from multiple theoretical perspectives, including sociocultural, sociolinguistic, attitudinal/motivational, cognitive science, and critical theory. Despite this theoretical variety, the studies covered herein share the commonality of focusing on the racial/ethnic, cultural, linguistic, and socioeconomic contexts of student diversity in science education.

Accountability as the Policy Context for Science Education

Research endeavors as well as schooling itself are influenced to some degree by policy – not only educational policy but also language policy, immigration policy, and other policies affecting children's and families' access to various social goods. Since it is beyond the scope of the present work to address all of these areas, the following discussion is limited to educational policy.

Under the current education reform focusing on standards-based instruction, content standards or curriculum frameworks at the national and state levels offer guidelines for school curricula and classroom instruction (Cohen & Hill, 2000; Knapp, 1997; McLaughlin et al., 1995; Smith & O'Day, 1991). Adherence to content standards may not always promote "best practices" in science education. For example, an emphasis on discrete facts or basic skills discourages teachers from promoting deeper understanding of key concepts or inquiry practices (Bianchini & Kelly, 2003).

Within the current context of high-stakes assessment and accountability, science has a unique status compared to core subjects such as reading, writing, and mathematics. Overall, in today's school curricula, science is emphasized far less than these other subjects (Hewson et al., 2001;

---

[2] As mentioned earlier, this book does not address the literature on science achievement and gender. However, beyond the issues of "gender gaps" in science, feminist scholars have offered critical epistemological challenges to Western modern science. None of the studies reviewed in this book took an explicitly feminist perspective; however, interested readers are referred to Keller and Longino (1996) and the previously mentioned works by Haraway (1990, 1991).

Knapp & Plecki, 2001; Spillane et al., 2001). As states increasingly turn to standardized assessments for accountability purposes, what gets taught in the classroom is largely determined by what gets tested. Approximately 30 states currently administer statewide assessments in science (Council of State Science Supervisors, n.d.).[3] However, science assessment is often not part of accountability measures. When it is part of accountability, science assessment is usually required only at one specific grade in elementary, middle, or high school, rather than over the entire grade span (as is generally the case with reading and mathematics).

Since science is seldom included in accountability measures, it is taught to a minimal degree in the elementary grades and tends to be undersupported in the secondary grades, compared to core subjects of reading, writing, and mathematics. Resources and opportunities in science education for teachers, students, and families also tend to be limited (Hewson et al., 2001; Knapp & Plecki, 2001; Spillane et al., 2001). For example, development of science instructional materials is not in high demand, science supplies and equipment are scarce or not easily accessible, neither large-scale nor classroom assessment instruments are widely available, and professional development opportunities are generally inadequate.

Similarly, research and development efforts in science education do not receive high priority. At the elementary level, schools rarely have science textbook series that are integrated across the entire elementary grade span, and science curricula tend to be discretionary and variable (Kennedy, 1998). Science textbooks and instructional materials are often of poor quality, and do not generally come with assessment instruments. As a result, science education researchers often must develop the instructional materials or assessment instruments necessary to carry out their research (Kennedy, 1998). This requires researchers to have expertise in curriculum development and instrument design, in addition to the other types of expertise already required to conduct research. It also means dedicating a substantial amount of a project's resources to the development, refinement, and pilot testing of instructional materials and/or assessment instruments.

The negative impact of educational policies affecting science education tends to be greater for nonmainstream students and students in inner-city schools.[4] For example, in states requiring accountability in literacy

---

[3] The use of statewide science assessment is spreading to more states, and this is likely to intensify as science will be part of the No Child Left Behind Act from 2007. For this reason, the exact number of states that assess science and/or include science in accountability measures changes over time. Furthermore, different documents present contradictory information in this regard.

[4] Other state and federal policies that indirectly affect education, such as those concerning migration, low-wage labor, and social services, also tend to affect nonmainstream students disproportionately.

and mathematics but not in science, the pressure for accountability overshadows the concern for elementary students' access to high-quality learning opportunities in science. Science instruction for students in inner-city schools (who are disproportionately low-income or ELL students) is often deemphasized relative to the urgent task of developing English proficiency as well as basic skills in literacy and numeracy (Lee, 1999a; Lee & Avalos, 2002). Assessment accommodations for ELL students in large-scale science assessments are either not considered or not consistently implemented, resulting in imprecise knowledge about the strengths, needs, and academic progress of these students (Abedi, 2004; Abedi, Hofstetter, & Lord, 2004; Solano-Flores, & Trumbull, 2003). If high-quality instructional materials that meet current science education standards are difficult to find (NSF, 1996), materials that also take into account the cultural and linguistic diversity of today's classrooms are even scarcer (NSF, 1998). Thus, educational policies, especially accountability measures, influence educational practices; this in turn influences the amount and kind of research that can be conducted.

Language programs for ELL students have been a topic of debate among politicians and the public as well as educators (García & Curry Rodríguez, 2000; Wiley & Wright, 2004). Policies mandating different types of bilingual/ESOL education largely determine how subject areas are taught to ELL students. In states that support bilingual education, science instruction can build on students' prior knowledge in science and the home language while students develop English proficiency (Kelly & Breton, 2001; Rosebery, Warren, & Contant, 1992). Currently, more states are shifting from bilingual education to "English only" policies that disregard development of students' home language and fail to consider students' proficiencies in the home language as relevant to academic achievement (Gutiérrez et al., 2002). In these states, science instruction for most ELL students is conducted in English; thus, students must learn new academic content in a language that they are still in the beginning stages of acquiring. In addition, some students may be removed from their classrooms during science instruction to receive instruction for English language development, and thus may receive little or no science instruction until they are assessed as English proficient. Furthermore, they may be deemed English proficient well before they have mastered the academic register of English. All of these policies tend to restrict the science learning opportunities available to ELL students.

After almost a decade of accountability in reading, writing, and mathematics, many states are now moving to incorporate science in accountability measures. This trend coincides with the planned federal policy on science accountability within the No Child Left Behind (NCLB) Act of 2001 (Public Law No. 107–110, 115 Stat. 1425), (2002), scheduled to take effect

in 2007. These policy changes at the federal and state levels may bring
about dramatic changes in many aspects of science education. Given that
the NCLB Act is inadequately funded, it does not provide schools with the
resources necessary to meet the accountability standards it imposes. Thus,
the consequences of such changes will likely be greater for students at
underfunded inner-city schools (McNeil, 2000) and ELL students (Abedi,
2004) than for their mainstream counterparts.

STUDENT LEARNING AND CLASSROOM PRACTICES

In an attempt to address the gaps in science outcomes (broadly defined), this synthesis highlights factors related to students' race/ethnicity, culture, language, and social class. Some studies examine relationships between these factors and student outcomes; particularly, experimental studies offer causal explanations of the impact of treatments or interventions on science outcomes. Others explicate underlying processes involving student diversity in science teaching and learning. In this section, we present the results of research focusing on classroom practices: (a) student characteristics and science learning, (b) science curriculum (including computer technology), (c) science instruction, and (d) science assessment. Studies addressing bilingual or ELL students are discussed within each section. Some studies addressed multiple topics and thus are included in more than one area. In describing specific studies, we address research questions, theoretical perspective or conceptual framework, methods (e.g., research setting, participants, data collection and analysis), results, and conclusions.

3

Students and Science Learning

There is a rather extensive body of literature in the broad category of students and science learning, compared to more limited bodies of literature in other categories described later. The studies address a wide range of topics, frame issues from multiple theoretical perspectives, and use various research methods. Some studies examine student characteristics or beliefs related to science learning, others focus on learning processes within the context of science instruction, and yet others provide data on students' learning outcomes. Studies on science learning were often conducted in the context of instructional interventions; those studies that specifically focus on student learning are discussed here, whereas those focusing on teaching are discussed under "Science Instruction" in Chapter 5. Studies on students and science learning have addressed the following topics: (a) student characteristics and school experiences as they relate to science achievement and career choices, (b) students' cultural beliefs and practices in relation to science learning, (c) scientific reasoning and argumentation, (d) the sociopolitical process of learning, and (e) science learning among ELL students.

Factors Related to Science Achievement and Career Choice

In a comprehensive literature review, J. Oakes (1990) described a range of cognitive and affective attributes, school experiences, and societal influences that are linked to differences in science achievement, persistence, and career choice among racial/ethnic and gender groups. The studies reviewed in this section examined various combinations of student characteristics and school experiences.[1] These studies generally employed correlational methods in the analysis of national databases, including the NAEP,

---

[1] Factors related to students' family and home environments are described in Chapter 9.

the National Education Longitudinal Study of 1988 (NELS:88), High School and Beyond, and the Scholastic Achievement Test. Because most of the studies were published after 1990, the present review may be considered an update of the literature review by Oakes (1990).

Science Achievement

One set of studies examines factors related to science achievement by race/ethnicity and gender. S. Rakow (1985b) examined the influence of the following six variables on students' inquiry skills (i.e., science process skills): general academic ability (defined in terms of students' self-reports of their high school grades), motivation, classroom environment, quality of instruction, quantity of instruction, and home environment. The study used a national stratified random sample of about 2,000 17-year-olds who were assessed as part of the modified version of the NAEP. Multiple regression analysis indicated different patterns for the two subgroups. For White students, general academic ability was a major predictor, whereas the other five variables taken together accounted for a smaller portion of the variance. For non-White students, ability was not a major predictor, and the other five variables taken together accounted for approximately the same amount of variance as with White students. In other words, this study found that non-White students' science process skills were not a reflection of their general academic ability, but it was unable to account for the variance in these students' science process skills.

Using a series of regression analyses of NELS:88, L. S. Hamilton and colleagues (1995) examined student and teacher factors related to science achievement scores of more than 5,000 students in 8th grade and then again in 10th grade. E. M. Nussbaum, Hamilton, and R. E. Snow (1997) expanded the previous study with close to 3,900 of these same students when they reached 12th grade. At the 10th and 12th grades, race/ethnicity and gender showed the strongest effects on (i.e., were the greatest predictors of) spatial-mechanical reasoning. In 10th grade, Black and Hispanic students performed lower than White students (but there was no difference between Asian and White students). In 12th grade, Black and Asian students performed lower than White students (but there was no difference between Hispanic and White students). In both grades, females performed lower than males. With regard to the effect of SES, the results differed between the two grade levels; in 10th grade, low-SES students performed lower than middle-SES students, but in 12th grade there was no difference. (Data about spatial-mechanical reasoning are not available for 8th grade.) At each grade level, different patterns of student and teacher factors were associated with each science factor (e.g., spatial-mechanical reasoning, quantitative science, scientific reasoning, etc.) by race/ethnicity, SES, and gender. The results suggest that achievement tests are multidimensional and that using psychologically meaningful subscores can enhance test validity and usefulness that are missed by analyses based on total scores alone.

P. A. Muller and colleagues (2001) used longitudinal data from the three waves (8th, 10th, and 12th grades) of NELS:88 and employed hierarchical linear modeling (HLM) to examine student and school factors related to science achievement and growth rates in precollege science by race/ethnicity and gender. The results indicated that students' SES (a composite variable consisting of parents' education, occupation, and family income) and previous science grades from grade 6 to grade 8 strongly and positively correlated to students' 8th-grade science achievement across all racial/ethnic and gender subgroups. The quantity of science units completed in high school (defined as students' total Carnegie units in science derived from transcript data) was the only consistent predictor of science growth rates across all racial/ethnic and gender subgroups. However, the relationships between student and school factors and science growth rates differed greatly across racial/ethnic and gender subgroups.

S. Peng and S. Hill (1994) used the NELS:88 data with a sample of about 6,500 minority students (i.e., African American, Hispanic, and Native American and Native Alaskan) representing the percentages of these groups in the general population. The students were classified into four achievement quartiles based on their science and mathematics test scores in the 8th grade, using the national norms of these tests. The study examined the influence of three sets of factors (i.e., student characteristics, school context and experience, and family resources and activities) differentiating high- and low-achieving minority students in science and mathematics. Certain student and school factors correlated with science achievement, regardless of cultural and ethnic backgrounds. High achievers were more likely to be in a high achievement group and a college preparatory program, suggesting that they received more rigorous academic training and gained enough competence to study higher-level courses. Low achievers, in contrast, were less likely to display persistent effort and active involvement in schoolwork, to be attentive in class, and to complete homework, suggesting that they did not spend enough time on learning tasks.

Science Careers
Another set of studies examines factors related to students' choice of science majors or science-related careers by race/ethnicity and gender. S. Maple and F. Stage (1991) used a national sample of close to 2,500 Black and White students drawn from the High School and Beyond database and analyzed the data using the LISREL model. The model included students' background characteristics (using seven variables) and an array of high school experiences (using six variables) to explain students' choice of a mathematics or science major in college. The dependent variable, field of

study, indicated whether or not the students had declared a mathematics or science major two years after high school graduation. Significant predictors across racial/ethnic and gender groups included the choice of a mathematical/scientific field of study as a high school sophomore, number of mathematics and science courses completed through the high school senior year, attitudes toward mathematics, and various parental factors. However, there were many differences across subgroups, and the model explained nearly twice as much variance for the Black male, Black female, and White male subgroups compared with the White female subgroup. This finding indicates that different subgroups follow different paths in arriving at the selection of a mathematics/science major.

O. W. Hill, C. Pettus, and B. A. Hedin (1990) examined seven factors thought to influence science career choices: teacher/counselor encouragement, participation in science-related hobbies and activities, academic self-image, science-related career interest, parental encouragement and support, perceived relevance of mathematics and science, and mathematics and science ability. Although it is unclear in the article how ability was defined, it seemed to indicate primarily mathematics and science skills and knowledge (1990, p. 292). Critical thinking ability, especially, was emphasized as an ability subscale. The study used a series of three surveys with more than 500 middle and high school students in Virginia and employed a multivariate analysis of covariance (MANCOVA) focusing on the effects of race/ethnicity, gender, and personal acquaintance with a scientist or mathematician, with age as a covariate. The results showed that whereas Black students had significantly higher science-related career preference scores than their White counterparts, they scored significantly lower on the measure of critical thinking ability. The results also showed that across both races and for both male and female students, the major factor affecting their preferences for science-related careers appeared to be personal contact with a scientist, which suggests the importance of role models (particularly those from one's own race and gender) in choosing a science career.

J. Grandy (1998) used the LISREL model with data from a longitudinal survey of high-ability minority SAT takers to determine the effects of students' background and school variables on persistence in science over five years. Data collection occurred at three points in time over the five-year period, each with winnowing of the sample. The first data collection in 1985 involved high school students who had scored at least 550 on the SAT mathematics test section (which placed them among the top 29% of the SAT population) and had indicated that they planned to major in a science/engineering field in college. The second data collection occurred in 1987, two years after these students graduated from high school, by means of a lengthy questionnaire that was sent to all the students from the first pool. The third data collection occurred in 1990, five years after the students graduated from high school; students from the second pool were contacted by phone or a short questionnaire was sent to them. Eventually, the number of students for whom all the data were available totaled more than 2,500 (43% of the original sample). This final sample included students who persisted in science/engineering, as well as those who switched majors or dropped out of college after the sophomore year. The results indicate two key predictors for persistence in science from high school through college. The first involved commitment to science (defined in terms of the extent to which students had found a science/engineering field to which they could make a commitment and the extent to which they enjoyed their chosen major field) during the college sophomore year. The second involved the availability of minority support systems (defined in terms of the extent to which students had minority or female role models and advisors, the extent to which they had advice and support from advanced students of their own racial/ethnic group, and the extent to which they had a dedicated minority relations staff). The availability of role models and advisors of the same racial/ethnic background was shown to be especially important in building enthusiasm for science during the first two years of college.

Despite the small number of studies on factors related to science achievement and choice of science majors or science-related careers by race/ethnicity and gender, several observations can be made about their overall results. First, of various factors reported in the literature, three sets seem to play key roles: (a) critical thinking and reasoning; (b) prior academic training and achievement in science, such as science grades and course taking; and (c) support systems, including role models of similar racial/ethnic and gender background. Second, little is known about the variance in science outcomes by race/ethnicity and gender, and even less is known about nonmainstream students specifically. Third, although some factors are consistently significant across all racial/ethnic and gender subgroups, other factors play differential roles with specific subgroups. Finally, the results consistently highlight the need to disaggregate data by demographic subgroups. Disaggregated results could help researchers develop more valid and reliable predictors and models of science outcomes tailored to specific subgroups. These results would also help guide policymakers and practitioners in designing better educational interventions to effect changes in science outcomes across subgroups.

Cultural Beliefs and Practices

All students have developed knowledge, values, and ways of looking at the world by virtue of their primary socialization and participation in a specific cultural community. In considering equity in science learning, it is important to take into account the intellectual and other resources that diverse student groups bring to the science classroom, even though these may not be easily recognized by the mainstream (Lee & Fradd, 1998; Moje et al., 2001; Warren et al., 2001). Recognition of students' strengths and limitations in learning science enables them to learn high-status knowledge valued in science disciplines and school science, while valuing cultural knowledge and beliefs.

An emerging body of literature indicates that students from some cultural communities display beliefs and practices that are discontinuous with Western science as it is practiced in the science community and taught in school. From cross-cultural and multicultural education perspectives, equitable science-learning opportunities are those that allow students to successfully participate in Western science, while also engaging the knowledge, beliefs, and modes of discovery characteristic of their communities of origin (Aikenhead, 2001a; Aikenhead & Jegede, 1999; Snively & Corsiglia, 2001). In this way, students gain access to the high-status knowledge of Western science without being forced to choose between school success and membership in their own cultural group. A perspective that recognizes and values diverse views of the natural world and diverse ways of knowing opens possibilities for academic achievement, as well as the strengthening of students' cultural and linguistic identities.

Three main bodies of literature examine nonmainstream students' cultural beliefs and practices and the influences of such beliefs and practices on science learning. These literatures focus on: (a) worldviews, (b) communication and interaction patterns, and (c) making the transition between students' cultural beliefs and practices and the culture of Western science.

Worldviews

A large body of studies has examined worldviews of diverse groups of students within the United States and around the world (see Cobern, 1991, 1996 for conceptual underpinnings of worldviews as related to science and science education). Studies within the United States have focused on Native American students of the Traditional Kickapoo Band (Allen & Crawley, 1998), Yup'ik (or Yupiaq) students in southwestern Alaska (Kawagley, Norris-Tull, & Norris-Tull, 1998), Mexican American students (Klein, 1982), and African Americans and Hispanics as well as mainstream White students (Lee, 1999b).

A substantial body of literature concerns the worldviews of students in countries other than the United States as they relate to science education. These include studies of elementary and secondary students from the First Nations (Native Americans) in Canada (Snively, 1990; Sutherland & Dennick, 2002); the People's Republic of China (Gao, 1998; Gao & Watkins, 2002); Taiwan (Chin-Chung, 2001); Botswana (Kesamang & Taiwo, 2002; Prophet, 1990; Prophet & Powell, 1993); Nigeria (Akatugba & Wallace, 1999; Jegede & Okebukola, 1991a, 1991b, 1992; Ogunniyi, 1987, 1988; Okebukola & Jegede, 1990); South Africa (Hewson, 1988; Lemmer,

Lemmer, & Smit, 2003); Zimbabwe (Shumba, 1999); Malawi (Dzama & Osborne, 1999); the Maori of New Zealand (McKinley, 2004; McKinley, Waiti, & Bell, 1992); Melanesia (Waldrip & Taylor, 1999a, 1999b); the Philippines (Arellano et al., 2001; P. P. Lynch, 1996a, 1996b); the Solomon Islands (Ninnes, 1994, 1995); and the West Indies (George, 1999). Some studies included preservice science teachers in Nigeria and the United States (Cobern, 1989) and in South Africa (Lawrenz & Gray, 1995), as well as practicing science teachers in Botswana, Indonesia, Japan, Nigeria, and the Philippines (Ogunniyi et al., 1995).

A majority of the studies, both in the United States and in other countries, used questionnaires or interviews to make inferences about students' worldviews. Field observations were rare. A small number of studies used experimental designs or correlational methods to examine the relationships of worldviews to contextual variables, such as residence patterns (e.g., rural, urban, suburban), family type (e.g., nuclear or extended), religion, literacy, exposure to science education, and gender.

Taken together, the studies on worldviews offer several key findings. First, the results indicate the shared and public acceptance of supernatural, spiritual, animistic, or volitional accounts of nature among students of nonmainstream backgrounds in the United States and in developing countries around the world. For example, Lee (1999b) examined children's views of the world after they personally experienced a major natural disaster (i.e., a hurricane). The study addressed three issues: (a) children's knowledge of the hurricane, (b) children's worldviews with regard to the causality of the hurricane, and (c) children's sources of information about the hurricane. The study involved more than 120 4th- and 5th-grade students in two elementary schools located in areas that were particularly hard hit by the hurricane. The sample included students by ethnicity and SES (and also gender). Both quantitative (ANOVA and frequencies) and qualitative methods were used.

The results indicate that students' interpretations of the event differed by ethnicity and SES. African American and Hispanic elementary students attributed the cause of the disaster to societal problems (e.g., racism, crime, violence) and spiritual and supernatural forces (e.g., God, the devil, evil spirits) more often than did Anglo students, who tended to give explanations in terms of natural phenomena. Likewise, low-SES students attributed the cause to wrongdoing by themselves or their family members (e.g., divorce, fights, drug use), societal problems, and spiritual or supernatural forces more often than did middle-SES students. This pattern of results was more pronounced when ethnicity and SES were examined together.

Anglo students were more knowledgeable about the scientific aspects of the hurricane than were African American and Hispanic students. Middle-SES students were more knowledgeable than low-SES students. When ethnicity, SES, and gender were examined together, male Anglo students from middle-SES backgrounds were the most knowledgeable of all groups. Students' worldviews were related to their science knowledge of the event – students who expressed alternative worldviews in their reasoning about the causes of the hurricane also tended to have incorrect science knowledge about the hurricane.

Although television and parents were the two most important sources of students' knowledge about the hurricane, there were notable differences among ethnic and SES groups. Anglo students generally obtained information from different sources that were relatively consistent with one another as well as with the view of Western science, whereas African American and Hispanic students obtained information from sources that were often incompatible with one another and the view of Western science. Low-SES students generally obtained limited or inaccurate information from parents as the most important source, whereas middle-SES students obtained scientifically based information from television as the most important source. Again, this pattern of results was more pronounced when ethnicity and SES were examined together.

A second key finding about the studies on worldviews, specifically those conducted in African countries, is that African students often expressed alternative worldviews even after taking science courses; that students with a high level of belief in traditional African worldviews did not perform in science as well as those with a low level of such beliefs; and that students who took more science courses expressed more scientific worldviews and had higher science achievement and more positive attitudes than those with fewer science courses.

A. H. Akatugba and J. Wallace (1999) examined why Nigerian high school students had difficulties using propositional reasoning to solve physics tasks. An interpretive case study was carried out with six physics students, selected through purposeful sampling, who could provide the broadest scope of information. The researchers related students' ability to use propositional reasoning to their sociocultural context, which included traditional African worldviews, students' discomfort with asking questions of teachers (as students perceived it to be disrespectful), and interpersonal teacher–student relationships (e.g., students perceived teachers' authoritarianism as discouragement from asking questions or being inquisitive).

F. Lawrenz and B. Gray (1995) examined the worldviews of 48 preservice science teachers (of various racial/ethnic, linguistic, and religious backgrounds) in a science methods class in South Africa. The study used chi-square tests to examine the relationships among worldviews, individual characteristics, and scientific concepts. The results showed that science was not well integrated into most of the preservice teachers' worldviews, although they had taken numerous science courses. The three characteristics that were most significantly related to participants' worldviews were type of courses taken, area of residence, and type of family situation. Especially preservice teachers who had taken more physics tended to have a more quantitative, orderly, and controlling view of the world.[2]

O. J. Jegede and P. A. Okebukola (1991a, 1991b, 1992) and Okebukola and Jegede (1990) examined the relationships between traditional African cosmology and science learning in Nigeria. They (1990) studied secondary school students to test the hypothesis that sociocultural variables influence students' attainment of science concepts. In a fixed-effect factorial design, the study tested the following variables: the general environment of the community (rural or automated), students' reasoning pattern (empirical or magical/superstitious), students' goal structure preference (cooperative, competitive, or individualistic), and nature of the home (authoritarian or permissive). The results showed that: (a) students who lived in predominantly automated environments did significantly better in concept attainment than those in predominantly rural environments; (b) students whose reasoning patterns were predominantly empirical did significantly better than those whose reasoning was more magical; (c) students who expressed preference for cooperative learning did significantly better than those who expressed preference for competitive or individual work; and (d) students from permissive homes did significantly better than those from authoritarian homes. The researchers (1991b) also reported that Nigerian high school students with a high level of belief in African traditional cosmology made significantly fewer correct observations of natural phenomena in science classes, compared to those with a low level of such belief. Using a pre- and posttest experimental design, the researchers (1991a) found that Nigerian high school students, after participating in a six-week instructional intervention in which they evaluated African traditional beliefs in light of scientific knowledge acquired in inquiry-based biology lessons, had higher science achievement and more positive attitudes toward science than did those in the control group.

A third key finding about the studies on worldviews is that although the conducting of scientific inquiry is a challenge for most students, it presents additional challenges for students from cultures that do not encourage them to engage in inquiry practices, such as asking empirical questions about natural phenomena, designing and implementing systematic investigations, and finding answers on their own. The studies on worldviews mostly focused on cultures whose traditional child-rearing and/or formal

---

[2] Although this study included subjects of various racial/ethnic, linguistic, and religious backgrounds, analysis was not conducted with regard to these variables. Given South Africa's history of racial segregation and the persistent tendency of settlement patterns to follow racial/ethnic lines, one might assume that "area of residence" (or even "type of family situation") could be read as an indicator of subjects' racial/ethnic, linguistic, or religious background. However, the authors of the study do not address this question.

educational practices are authoritarian in nature (e.g., Akatugba & Wallace, 1999; Arellano et al., 2001; McKinley et al., 1992; Ninnes, 1994, 1995; Prophet, 1990; Shumba, 1999; Waldrip & Taylor, 1999a, 1999b). In these societies, cultural norms prioritize respect for teachers and other adults as authoritative sources of knowledge, rather than the development of theories and arguments based on students' own evidence and reasoning. Children are taught to respect the wisdom and authority of their elders and are not encouraged to question received knowledge. As inquiry and critical questioning are generally not encouraged at home or in school, verbalization by children of their ideas and discussion about why certain answers were correct or incorrect were rarely observed in science classrooms.

The study by N. J. Allen and F. E. Crawley (1998) is unique in several respects: (a) it is one of a small number of worldview studies conducted in the United States; (b) it used ethnographic methods (interviewing, participant observation) over an extensive fieldwork period; (c) it examined multiple perspectives of participants, including students, teachers who shared the students' background and those who did not, and community members; (d) it was conducted in both science classrooms and community settings; and (e) it highlighted points of congruence between participants' traditional views and modern scientific perspectives, whereas most other studies have focused on points of conflict between the two. The researchers report differences between the worldview of middle school students and adults of the Traditional Kickapoo Band of the southwestern United States and the worldview expressed during science instruction by two teachers who did not share the students' background. Differences were observed with regard to epistemology, preferred methods of teaching and learning, spatial/temporal orientation, behavioral norms, and subjects' perspective on the place of humans in the natural world. However, the researchers noted that some of the pedagogical methods preferred by the Kickapoo students (e.g., cooperative learning and holistic content) are highly effective, and that their views of nature are in tune with modern ecological perspectives. The researchers argue that although none of the worldview differences they noted would directly prevent the students from being full participants in the scientific community, many of these differences seem to prevent them from being successful in the science classroom, due to the science teachers' failure to see the connections between students' worldviews and scientific beliefs and practices.

Communication and Interaction Patterns
Research has examined culturally specific communication and interaction patterns with regard to various aspects of science learning among students from diverse backgrounds (see the review by Atwater, 1994). Literature reviews have addressed science education among African American (Atwater, 2000; Norman et al., 2001), Asian American (Lee, 1996),

Hispanic (Rakow & Bermudez, 1993), and Native American students (Kawagley et al., 1998; Nelson-Barber & Estrin, 1995, 1996; Riggs, 2005). These reviews usually indicate that cultural norms and expectations affect science learning within each group and that these are often inconsistent with the expectations of school and school science. Based on such results, the reviews draw implications for science learning and highlight the importance of incorporating students' cultural patterns as a knowledge base for culturally congruent instruction (discussed in the "Science Instruction" chapter).

A small number of studies have examined culturally specific communication and interaction patterns within the context of science instruction. Since the early 1990s, O. Lee and colleagues have conducted a programmatic line of research to promote science learning and English language development among elementary students from diverse linguistic and cultural backgrounds (see the details in Lee, 2002). At the start of this research, Lee and S. H. Fradd (Lee & Fradd, 1996a, 1996b; Lee, Fradd, & Sutman, 1995) worked with dyads of Hispanic, Haitian, Anglo, and African American elementary students. The students interacted with teachers who were matched in terms of ethnicity, language, and gender (e.g., a dyad of Haitian girls with a Haitian female teacher) while working on science tasks outside the classroom setting. The eight participating teachers were selected for their expertise in working with nonmainstream students, and most had advanced degrees in ESOL/bilingual education. Interviews were conducted in the languages of students' choice (English, Haitian Creole, and/or Spanish). Given the small number of students in the entire group and especially in subgroups, data were analyzed in terms of descriptive statistics (i.e., means of scores or frequencies of responses), as well as major patterns and themes emerging from qualitative analysis. The results from this series of studies, described here, indicate similarities and differences with regard to communication and interaction patterns among culturally and linguistically diverse groups of elementary students.

Lee, Fradd, and F. X. Sutman (1995) examined science vocabulary, science knowledge, and cognitive strategy use among the student groups. Compared to Anglo students, students from nonmainstream backgrounds had more difficulty with science knowledge and vocabulary. Some did not have personal experience or prior knowledge related to the science tasks under study (e.g., swimming as an example of buoyancy or playing on a seesaw as an example of lever). Others had an understanding of the science concepts but lacked the specific vocabulary to convey precise meanings. Still others had minimal prior schooling, and therefore lacked exposure to science learning environments as well as science knowledge and vocabulary. Furthermore, students from different backgrounds used different kinds of cognitive strategies while engaging in the science tasks. Despite their difficulties, nonmainstream students also demonstrated strengths, such as improved performance when science tasks related to their prior experiences, a desire to engage in science activities, and positive feelings about the teachers.

Lee and Fradd (1996a) examined students' understanding and production of written and pictorial representations of science concepts. Specifically, the study focused on two aspects of literacy related to science performance: (a) students' interpretations of pictorial representations depicting a series of science activities and (b) students' written summaries of science activities. The results indicate that students generally had difficulty interpreting pictorial representations of science activities and communicating their ideas about science in written forms. The results also indicate differences among student groups in terms of the quantity and quality of written and pictorial communication, as well as in their enjoyment and appreciation of science activities. The researchers argue that since representations of scientific ideas often occur in written and pictorial forms, science instruction needs to promote science learning and literacy development simultaneously.

Lee and Fradd (1996b) examined patterns of verbal discourse, nonverbal communication, and engagement in science tasks among student-teacher triads during science tasks. The results indicate consistent patterns within each group but distinct differences among the groups. Consistent patterns of verbal discourse and nonverbal communication were observed with each group with regard to talk (linear or overlapping), turn taking (sequential or simultaneous), unit of discourse (complete sentences or words/phrases), and nonverbal communication (gestures and facial expressions). Additionally, consistent patterns of task engagement were observed within each group with regard to task performance (step-by-step or overlapping activities), mode of teacher guidance (probing/eliciting or telling/teaching), teacher reinforcement (information feedback or social, motivational statements), and student initiative (task-related or social-personal). Consistent with the literature on cultural congruence, teachers and students of the same cultural and linguistic background interacted in ways that promoted students' participation and engagement.

According to J. Lipka (1998), Yup'ik children in Alaska learn science-related skills (e.g., fishing, building fish racks, and using stars to navigate) from observing experienced adults and actively participating as apprentice-helpers in home and community settings. Children and adults engage in joint activities for long periods of time, during which verbal interaction is not central to the learning process. This style may not be optimal within a traditional Western school system that organizes learning around short and frequent class periods in which students are expected to listen passively to teachers, follow directions, and respond verbally to questions. The communication and interaction patterns in the Yup'ik community are grounded in spiritual, pragmatic, and inductive processes of thinking in this culture (Kawagley et al., 1998). Natural phenomena are explained in terms of readily observable characteristics or experiences involving a high degree of intuitive thought. Many of the skills and values that are instilled through the culture (e.g., participation and interaction with nature, a contextualized and personalized approach to knowledge) are useful in teaching and learning science. In contrast, Western scientific approaches to the natural world (e.g., an analytical and depersonalized approach to knowledge) can be alienating for Yup'ik youth.

Cultural Transition

The literature on worldviews and communication and interaction patterns in relation to science learning, as discussed, indicates that the culture of Western science is foreign to many students (both mainstream and non-mainstream), and that the challenges of science learning may be greater for students whose cultural, epistemological, and discursive traditions are discontinuous with the ways of knowing characteristic of Western science and school science. The challenge for these students is "to study a Western scientific way of knowing and at the same time respect and access the ideas, beliefs, and values of non-Western cultures" (Snively & Corsiglia, 2001, p. 24).

The ability to shift between different cultural contexts is critically important to nonmainstream students' academic success. H. Giroux (1992), among others, has used the notion of *border crossing* to describe this process. To succeed academically, nonmainstream students must learn to negotiate the boundaries that separate their own cultural environments from the culture of Western science and school science (Aikenhead, 2001a; Aikenhead & Jegede, 1999; Costa, 1995; Jegede & Aikenhead, 1999). For example, V. B. Costa (1995) studied high school students enrolled in chemistry or earth science in two schools with diverse student populations. She identified five types of students based on the relationships between their home worlds and the worlds of school and science (1995, p. 316):

- Potential Scientists: Worlds of family and friends are congruent with worlds of both school and science.
- "Other Smart Kids": Worlds of family and friends are congruent with world of school but inconsistent with world of science.
- "I Don't Know" Students: Worlds of family and friends are inconsistent with worlds of both school and science.
- Outsiders: Worlds of family and friends are discordant with worlds of both school and science.
- Inside Outsiders: Worlds of family and friends are irreconcilable with world of school, but are potentially compatible with world of science.

Even within the cultural mainstream, relatively few children's primary socialization is so "science-oriented" as to be perfectly continuous with the demands of school science. Thus, border crossing between the culture of Western science and the culture of the everyday world is demanding for all students (Driver et al., 1994; O'Loughlin, 1992). At times, students may find themselves torn between what is expected of them in science classes and what they experience at home and in their community. If they appear too eager or willing to engage in science inquiry, they may find themselves estranged from their family or peers. If they appear reluctant to participate, they risk marginalization from school and subsequent loss of access to learning opportunities. Although some students may successfully bridge the cultural divide between home and school, others may become alienated and even actively resist learning science.

Scientific Reasoning and Argumentation

In contrast to the literature indicating that students from nonmainstream backgrounds display ways of knowing that are sometimes discontinuous with Western science or school science, described earlier, an emerging body of literature argues that the ways of knowing and talking characteristic of children from outside the cultural and linguistic mainstream are generally continuous with those characteristic of scientific communities. Drawing on both a cognitive science perspective and the sociology of science, this research primarily employs discourse analysis of students' oral and written communication as they interact with teachers or peers during scientific inquiry tasks.

Based on detailed analyses of the everyday practice and talk of scientists, recent work in the sociology of science defines science and scientific practices more broadly than traditional definitions that emphasize experimentation and theory building (Latour & Woolgar, 1986; M. Lynch, 1985). This expanded view considers scientific practices to be embedded within the personal, social, and historical contexts of scientific communities. It also considers the role of imagination, conjecture, "cultivation of the unexpected," beliefs and desires of individual scientists, and construction of variables during the process of investigation, rather than control of predetermined variables.

Based on this expanded view of science and on a more flexible and fluid view of children's everyday sense making, A. S. Rosebery, B. Warren, and their colleagues on the Chèche Konnen Project have examined the complex, interactive, and complementary relationships between scientific practices and the everyday sense making of children from diverse cultures and languages (Ballenger, 1997; Rosebery et al., 1992; Warren et al., 2001; Warren & Rosebery, 1996; Warren, Rosebery, & Conant, 1994). Following a programmatic line of research since the late 1980s, the Chèche Konnen team has conducted case studies of low-income students from African American, Haitian, and Latino backgrounds in bilingual and regular classrooms (see the details in Lee, 2002). It highlights the continuity between the forms of reasoning and argumentation characteristic of nonmainstream, low-income students and those characteristic of scientific communities. It also highlights how the students draw upon their everyday knowledge when engaged in scientific inquiry, reasoning, and argumentation. Warren, C. Ballenger, and colleagues (2001) state:

> Traditionally, those who have thought about the relationship between particular cultural groups and the culture of science have identified tensions between what they describe as the knowledge, values, and practices of science and the knowledge, values, and practices of children from particular racial, ethnic, and linguistic minority communities. We would argue that the perspective we have put forward in this article can effectively reframe these tensions by opening up for examination what is meant by science on the one hand and diversity in cultural and linguistic practices on the other. By examining both in an integrated and reflexive way, we can begin to envision pedagogical possibilities that build on diversity as an intellectual resource rather than a problem or tension in science learning. (p. 548)

The results indicate how these students' ways of knowing and talking are continuous with those of scientific communities (Ballenger, 1997; Rosebery et al., 1992; Warren et al., 2001; Warren & Rosebery, 1996; Warren et al., 1994). The results also indicate that these students deploy sense-making practices – deep questions, vigorous argumentation, situated guesswork, embedded imagining, multiple perspectives, and innovative uses of everyday words to construct new meanings – that serve as intellectual resources in science learning. For example, while Haitian students are typically quiet and respectful in the classroom, in a culturally familiar environment they can participate in animated arguments about scientific phenomena in a way that is integral to Haitian culture and congruent with scientific practice (Ballenger, 1997). Students as young as 1st grade employ accounts of everyday experiences, not merely as a context for understanding scientific phenomena but as a perspective through which to infer previously unnoticed aspects of a given phenomenon and create possibilities for interpreting the phenomenon differently. Warren, Ballenger, and colleagues (2001) conclude:

> We are arguing for the need to analyze carefully, on one hand, the ways of knowing and talking that comprise everyday life within linguistic and ethnic minority communities and, on the other hand, the ways of talking and knowing characteristic of scientific disciplines.... What children from low income, linguistic, and ethnic minority communities do as they make sense of the world – while perhaps different in some respects from what European American children are socialized to do – is in fact intellectually rigorous and generatively connected with academic disciplinary knowledge and practice. (p. 546)

Students' scientific inquiry, reasoning, and argumentation occur in the context of social group dynamics. Individuals' racial/ethnic, cultural, linguistic, and social class backgrounds shape their interactions and communication patterns, which in turn limit or foster opportunities to engage in scientific practices. Looking closely at small-group interactions among four middle school students of different races and social classes, L. A. Kurth, C. W. Anderson, and A. S. Palincsar (2002) found that the students often failed to achieve intersubjective communication about the science task during group work. In particular, an African-American girl was unable to hold the floor within the group, and her opportunities for science learning were diminished. These failures occurred despite the fact that the curriculum materials and the teacher aimed to promote students' reasoning and argumentation, and that the students all wanted to share ideas and understand the phenomenon in question. The expectations that the students brought with them about how and when people should talk, how work should be done, and the standards of quality to which they should aspire led them to construct troubling inequities among themselves. The researchers interpreted the results in terms of how the children's actions were connected to their family histories, how the privileging of ideas was connected to privileging of people, and how the practice of science was connected to discrimination. They also point out the difficulty of separating conceptual conflict about scientific ideas from interpersonal conflict about privilege and status.

The Sociopolitical Process of Science Learning

As an outgrowth of critical studies of schooling, a small number of studies have examined science learning as a sociopolitical process. This literature has several features that distinguish itself from other areas of research on science learning. First, it questions the relevance of science to students who have traditionally been underserved by the education system, and argues that science education should begin with the intellectual capital of the learner and his/her lived experiences, not with externally imposed standards. In this way, it attempts to invert the power structure of schooling and its oppressive effects on students from marginalized groups.

Second, this literature addresses issues of poverty, in addition to cultural and linguistic diversity, from a critical perspective that focuses on the unequal distribution of social resources. Researchers in this tradition generally employ critical ethnography (which combines the theoretical frame of critical theory and the methodology of ethnography) and ground their analyses in the political, cultural, and socioeconomic history of the groups under study.

Finally, this approach "allows and necessitates that educators make a political commitment to the struggle for liberation of the oppressed and thus provides a bridge between research and activism" (Seiler, 2001, p. 1003). The boundary between research and scholarship, on one hand, and advocacy and activism, on the other hand, is purposely blurred in this literature.

While most critical ethnography of schooling touches upon various school subjects to address more general processes of socialization and stratification within schools, a few studies have focused specifically on science learning. A. J. Rodriguez and C. Berryman (2002) worked with 38 10th-grade students in predominantly Latino and impoverished school settings in a U.S.-Mexican border city. The instructional approach in the study was guided by a "sociotransformative constructivism" that merged multicultural education with social constructivism. Using a curriculum unit on investigating water quality in their community, the students engaged in authentic activities as they explored how this topic was socially relevant and connected to their everyday lives. The researchers asked students to complete a semistructured concept map on the topic before and after instruction, and analyzed pre- and postinstruction concept map scores using a dependent t-test. The quantitative results indicate that the instructional approach enhanced not only students' enthusiasm for the science curriculum but also their knowledge and understanding of science content. Additionally, qualitative analyses indicate that students took on empowered positions by testing water in their homes and investigating ways to improve water standards in their communities. They quickly understood the precariousness of water availability in their desert region and informed their families of ways to conserve water at home. Having come to see science as relevant to their lives, students saw scientific investigations as worthwhile for themselves and for students in other schools in the region. The researchers conclude that their approach has the potential to open empowering spaces where students can engage with science curricula in socially relevant and transformative ways.

Recognizing the persistent achievement gap between mainstream White and inner-city African American students, G. Seiler (2001) used critical ethnography to describe a science lunch group that she organized with eight African American male students in an inner-city high school. The students and the researcher met once each week to eat lunch, talk about their lives, and discuss and carry out science-related activities. In this informal setting within the school, science activities started from students' own interests, prior knowledge, and lived experiences. The results indicate how the science lunch group forged a learning community based on respect and caring that afforded these African American teens the opportunity to participate in science in new ways. The researcher argues that the imposition of external standards on inner-city schools does little to ameliorate achievement gaps because it fails to address the significance of students' social and cultural lives. According to Seiler, critical ethnographic research could enable science educators to learn from students about how science education can change to meet students' aims and interests.

A. Calabrese Barton (1998a, 1998b, 1998c, 2001) has carried out research with children living in poverty, specifically urban homeless children living in shelters. She argues that to make science relevant to these students, not only students but also science itself must change, and that the reflexive nature of the relationship between science and students' lives should be made explicit. In a series of critical ethnographic studies, she presents narratives of how urban homeless children make sense of science on the basis of their lived experiences. She argues that educational research, including science education research, is centrally about making a political commitment to struggles for liberation. (More detailed results of this research program are presented in Chapter 9).

Overall, the studies described here reject the notion of imposing external standards on students who have been marginalized from science and science education. They argue instead that these students should be guided to make sense of science on the basis of their lived experiences in social and cultural contexts. The results indicate that this approach can lead students to gain knowledge and understanding of science content, see science as relevant to their lives, and engage in science in socially relevant and transformative ways.

Science Learning among ELL Students

A number of studies focus on linguistic influences on the science learning of ELL students in either bilingual or mainstreamed classrooms. These studies use a variety of methods, ranging from discourse analysis of students' elicited responses about science concepts to statistical analysis of multiple factors affecting students' performance on standardized science assessments. Inasmuch as they attempt to examine a broad range of cultural and linguistic diversity, many of these studies have been conducted outside the United States, in other parts of the English-speaking world. Due to the wide range of theoretical, methodological, and policy perspectives represented, it is difficult to draw coherent generalizations from this body of work. Nevertheless, most of the studies coincide in finding that students' limited proficiency in English constrains their science achievement when instruction and assessment are undertaken exclusively or predominantly in English. Studies undertaken within the United States did not posit a major instructional role for students' home languages; in contrast, those studies from countries in which language policies allow for greater presence of other languages in the classroom point to the cognitive and ideological importance of students' home languages and their usefulness for at least some academic functions.

Studies within the United States

Research on the science learning of ELL students within the United States has, not surprisingly, focused on Spanish speakers. The literature search on this topic produced two studies in the United States, which examined disparate research questions using very different research methods. In an interpretive study, B. J. Duran, T. Dugan, and R. Weffer (1998) studied how 14 Mexican American high school students constructed understandings of biology concepts based on their extant language skills in English. The students were enrolled for 32 weeks in a Saturday enrichment program at a private university that maintained a partnership with the public high school. They were taught by two teachers with extensive teaching experience, one in reading and the other in biology. (The biology teacher spoke only English, which suggests that the biology instruction was conducted exclusively in English, though this was not specified.) Initially, students were overly reliant on the biology teacher's talk as a source of science meaning, placed a high value on the science expertise of the teacher, and deferred to him as the sole scientific authority. Through instructional activities designed to show students how to use semiotic tools (i.e., diagrams) to construct and express conceptual meanings, students proceeded through three phases of learning. At the phase of *receptive* understanding and expression, they used diagrams to identify content and to "ventriloquate" or mimic the actions and speech of the teacher. At the phase of *conceptual* understanding and expression, they constructed conceptual meanings and advanced from "ventriloquating" teacher talk to expressing concepts for their own purposes. Finally, at the phase of *interpretative* understanding and expression, they used their conceptual understanding to analyze real-life experience and their expanded discursive resources to write their own interpretations. As students became more proficient with semiotic tools, they assumed responsibility for constructing meaning using their own discursive resources, and the teacher withdrew as the sole scientific authority. The results demonstrate the importance of providing language minority students with opportunities to acquire the discourse of science and other semiotic tools.

Grounding their work in J. Cummins's distinction between "basic interpersonal communicative skills" (BICS) and "cognitive academic language proficiency" (CALP) in second language development, H. N. Torres and D. L. Zeidler (2002) used a three-way factorial design to examine the effects of three independent variables (CALP in English, scientific reasoning skills, and "language learners") on the dependent variable (scientific content knowledge). CALP in English (low, intermediate, or high) was measured by the Test of English as a Foreign Language (TOEFL). Scientific reasoning (intuitive, transitional, and reflective) was measured by A. Lawson's instrument; the instrument was translated into Spanish for those with low or intermediate TOEFL scores. The "language learner" variable distinguished between Spanish-speaking ELL students and native English speakers. Science content knowledge was assessed using the 1999 Grade 10 Massachusetts Comprehensive Assessment System (MCAS) administered in English. The ANOVA results indicated that while the language learner variable did not have any statistically significant effect on students' performance on the MCAS, students' level of English language proficiency and their scientific reasoning skills both had significant effects, independently and in interaction with each other. The results suggest that combined high levels of English language proficiency and reasoning skills enhance students' abilities to learn scientific content knowledge in English.

Studies outside the United States
Research undertaken in other parts of the English-speaking world has focused on students from a broad range of language communities, both immigrant and indigenous. In contrast to studies undertaken in the United States, the following studies focus mainly on the role of ELL students' home language in learning science. Some of these studies go beyond examination of students' use of either the home language or English in the classroom to consider the social, cultural, and demographic dynamics of language communities (e.g., functions and attitudes associated with English and students' home languages both within and beyond the school).

K. Tobin and C. McRobbie (1996) conducted qualitative research on the ways in which eight ELL Chinese high school students in Australia endeavored to use English to make sense of what happened in class and to demonstrate the extent to which they had learned chemistry. The results indicate that the students employed Cantonese in their oral and written discourse and exhibited high levels of effort, commitment to learn, and task orientation both in and out of school. Students' work ethic was consistent both with the expectations of the teacher and with typical schooling practices in their home country. Despite the students' efforts to learn and understand chemistry, they were limited by their difficulties in English. The results suggest that a linguistic hegemony based on the use of English to teach chemistry and assess performance placed these ELL students in a position of potential academic failure. The researchers argue that learning can be facilitated when ELL students are provided with opportunities to fully employ their native language tools, when science instruction utilizes the cultural capital of the students, and when the microculture of the classroom fits the macroculture of students' lives outside the classroom.

J. Kearsey and S. Turner (1999) examined whether bilingual students had an advantage with regard to acquiring the specialized linguistic register of science, due to their broader experience with language learning and the linguistic awareness that this experience provides; or, whether interference between their two languages, combined with the additional "language load" implied by the scientific register, placed them at a disadvantage for science learning. The researchers evaluated the "accessibility" of a commonly used science textbook for bilingual students from Eastern Europe, Africa, Asia, and established immigrant communities and monolingual (English-speaking) students in 6 secondary schools in the United Kingdom. The study involved more than 200 students from 6 schools that were selected from a sample of 37 schools that had consented to participate. Data collection techniques included questionnaires, cloze tasks, readability formula analysis, and interviews. This study also considered the social contexts of bilingual students' language use, working from the concept of *diglossia* (i.e., functional differentiation of languages by bilingual speakers). The results in terms of frequencies of student responses indicate that for Gujarati-speaking students who were part of a diglossic community (i.e., who attended school with enough other Gujarati speakers to be able to use Gujarati regularly in contexts outside the home), bilingualism conferred an advantage over monolingual students with regard to performing close tasks and comprehending passages from the science textbook; this advantage was not evident in other bilingual students who were not members of diglossic communities. This advantage was ascribed to these students' greater linguistic awareness; however, it did not extend to tasks involving scientific writing, due perhaps to students' lack of familiarity with the complexities of English grammar. Bilingual students not living in diglossic communities were at a disadvantage compared with their monolingual (and bilingual Gujarati-speaking) peers. The researchers attribute this finding to linguistic interference between students' two languages,[3] and they note that it applied as much to economically advantaged students as to those from less-advantaged homes. They conclude that bilingual students could benefit from a range of curriculum materials supporting linguistic tasks of various levels of difficulty (in contrast to the standardized nature of the National Curriculum). They also note that the possible additive effects of bilingualism suggest that it should be treated as a resource in the classroom, and that a school culture that values bilingualism may foster an improved understanding of scientific language in bilingual pupils.

[3] It is unclear why Kearsey and Turner ascribe students' difficulties to their bilingualism, rather than to their lack of English proficiency. In fact, they end up concluding that bilingualism constituted a disadvantage for some students and an advantage for others. This finding suggests that the key factor is probably not bilingualism per se, and the evidence presented implies that living in a well-bounded diglossic community that stresses proficiency in both languages may be more important. In contrast to Curtis and Millar (1988), described next, Kearsey and Turner did not consider the length of time that students had spent in the British school system. Given the stability of their diglossic community, it seems likely that the Gujarati-speaking students had probably been in the UK longer than the other bilingual students.

Cognizant of the widespread perception that students of Asian background (i.e., from the Indian subcontinent) in the UK perform less well than mainstream British children, S. Curtis and R. Millar (1988) examined students' knowledge about, and associations with, basic scientific concepts among an unmatched sample of about 500 secondary students (ages 13+) from two schools. The study consisted of two groups of students: Asian students from homes where languages other than (or in addition to) English were spoken, and British students who were monolingual in English. Students were asked to write (in English) "All I know about . . ." six science terms (temperature, weight, speed, electric current, power, pressure), and their responses were coded for scientific content, accuracy, intelligibility, and relevance. Data were analyzed using dependent t-tests of mean scores between Asian and English-speaking students. The Asian students produced more "indecipherable" statements, suggesting that limited fluency in English affected their ability to express themselves clearly on the given task. The native English speakers gave more scientific ideas and applications, indicating their greater familiarity with the language of school science and of everyday situations related to science. (The researchers note that the monolingual students wrote more in general, including more incorrect statements, probably due to their greater fluency in English.) Students' length of school attendance in England was also shown to produce some statistically significant differences; in contrast to the "short stay" Asian students, the responses of Asian students with eight or more years of schooling in the UK were virtually indistinguishable from those of the native English speakers. The researchers conclude that the results did not indicate that science is any more difficult for Asian students, except insofar as language problems may hinder their learning and/or expression of ideas. They also stress that the value of such comparative studies may lie less in the information on differences between groups than in the more general insights concerning the role of language in the learning of all children.

In a series of studies involving students in Tasmania (Australia), the Philippines, and India, P. P. Lynch and colleagues (Lynch, 1996a, 1996b; Lynch, Chipman, & Pachaury, 1985a, 1985b) examined how students' mother tongue and degree of "Westernization" were associated with their understanding of science concepts. Lynch and colleagues (1985a) examined concept recognition between English-speaking high school students in Tasmania and their Hindi-speaking counterparts in the Bhopal/Barwani region of India. They found that in some cases, purely linguistic factors could aid in concept recognition. For example, the related Hindi terms *annu* ["atom"] and *pramanu* ["molecule"] themselves indicate the hierarchical relationship between atoms and molecules. Working from a Piagetian cognitive framework, the researchers (1985b) examined preferred thinking styles ("membership," "partial association," or "generalization," which the researchers also posited as differing levels of cognitive ability) between the same two student groups. Data were analyzed using chi-square tests of student responses between the two language groups. By the end of high school, a slight majority within each group preferred to think at the highest level ("generalization"), and developmental trends for the two groups were similar. However, the preference for generalization among the Hindi-speaking group was significantly lower than for the English-speaking group. Though the researchers did not explicitly state what might be the major cause of this result, linguistic factors and experience with different types of instruction were surmised to play only minor roles.

Lynch (1996a) examined how English-speaking Tasmanian students, Tagalog-speaking Philippine students, and B'laan-speaking Philippine students conceived of the Earth/Sun/Moon system, arguing that the scientific understandings of the Philippine students were constrained by both linguistic and cultural factors. Lynch (1996b) describes "cultural effects" with regard to the different criteria by which the Philippine students classified substances as solid, liquid, or gas, and also noted semantic differences between terms in students' own languages and the scientific conceptions of solid, liquid, and gas. Both studies were based primarily on data from 16 Tasmanian 3rd graders, 16 Tasmanian 6th graders, 20 Tagalog-speaking 6th graders, and 20 B'laan-speaking 6th graders. On the basis of the results of the two studies, Lynch interprets differences among the groups as evidence of students' "alternative frameworks." He argues that (a) "non-intellectualized" languages such as the indigenous languages of the Philippines are not, in their current form, adequate to correctly express scientific concepts (but still have an important educational role to play due to their cultural and ideological importance); and (b) quality science instruction for non-Westernized students necessarily involves reconstruction of students' worldview.[4]

Discussion

Research on student characteristics and learning in science has been conducted from a wide range of theoretical or conceptual perspectives,

---

[4] The work by Lynch and colleagues appears frankly assimilatory, inasmuch as it argues that scientific understanding and higher levels of cognitive ability are products of students' degree of Westernization. While it is logical that more Westernized students will perform better in a Western educational system, this narrow view of learning and cognition displays a remarkable disregard for the cultural and linguistic dislocation that forced assimilation (abetted by formal schooling) has imposed upon colonized peoples. This is not to argue that quality science instruction for indigenous students is anything less than urgent, but many researchers engaged in long-term work with indigenous communities will take issue with the suggestion that "reconstruction" of their worldview in a Western mold constitutes the solution.

including psychological, sociocultural, sociolinguistic, cross-cultural, cognitive science, and critical theory. Such range of perspectives portrays multiple facets of science learning among diverse student groups. Each perspective relies primarily on a particular research method, for example, correlational methods to examine student and teacher/school factors related to science achievement or career choice, surveys and interviews to examine worldviews, discourse analysis to examine students' scientific inquiry and reasoning, and critical ethnography to examine the sociopolitical aspect of science learning. These pairings of theory and method reflect traditions of educational research more broadly.

The extensive body of literature on student characteristics and science learning indicates that nonmainstream students' science learning is influenced by a variety of factors associated with their cultural, linguistic, and social-class backgrounds. These factors include students' cognitive and affective attributes, cultural beliefs and practices, cognitive processes underlying scientific inquiry and reasoning, and sociopolitical processes. With ELL students, the interplay between English and the home language is critical in learning science. Although it seems valid to conclude that all these factors contribute to nonmainstream students' science learning, it is difficult to specify the role of each factor, both independently and in interaction with the others, due to the limited literature within each area.

Results emerging from different research traditions are sometimes inconsistent or contradictory. This is probably due in part to differences in emphasis reflecting the conceptual, ideological, and political commitments among researchers. Research on cultural beliefs and practices and research on scientific reasoning and argumentation both emphasize the intellectual resources that nonmainstream students bring to science, and examine the intersections between students' linguistic and cultural experiences and scientific practices. However, the former addresses discontinuities between the prior cultural and linguistic knowledge of nonmainstream students and Western science, whereas the latter highlights continuity between these students' ways of knowing and talking and those characteristic of scientific communities. From the "discontinuity" perspective, science learning involves "border crossing" or making cultural transitions between the home culture and the culture of science. From the "continuity" perspective, nonmainstream students' ways of exploring ideas or investigating scientific questions already overlap considerably with the way science is practiced in scientific communities.

Inconsistent or contradictory results are also observed in the literature on science learning with ELL students. Some research on ELL students indicates additive effects of students' home language (Kearsey & Turner, 1999), whereas other research emphasizes the limitations of indigenous languages for purposes of science learning (e.g., P. P. Lynch, 1996a, 1996b).

In general, when instruction is in English, ELL students' science learning is in direct relation to their level of English proficiency.

The inconsistency of these results is probably due in part to the serious methodological limitations characterizing much of the literature on science learning among ELL students. Despite the focus on the linguistic mediation of students' science learning, few studies describe in any detail the language treatment that students received, making it difficult to interpret the findings. Furthermore, some studies seem to consider "ELL" as a fixed characteristic, failing to consider that progress in student science outcomes may reflect the fact that students' English proficiency has improved over the course of the research period. Generally speaking, the research base lacks the theoretical and methodological underpinnings necessary to make sense of ELL students' science learning. This underscores the urgent need for more cross-disciplinary research that can draw upon the insights of scholars with knowledge about different aspects of science learning.

4

Science Curriculum

The important role of curriculum materials in instructional reform has been emphasized by D. L. Ball and D. K. Cohen (1996), among others. Although curriculum development is a huge area of endeavor in science education, research on curriculum development, implementation, and effectiveness is relatively limited. Even studies that involve curriculum materials development as an essential tool for conducting the research often do not address the curriculum itself as a research topic.

While appropriate instructional materials are essential to effective instruction, high-quality materials that meet current science education standards are difficult to find, and are even less likely to be available in inner-city schools where nonmainstream students are concentrated. Despite this paucity of quality materials, especially for nonmainstream students, they are not in high demand compared to curriculum materials for the core subjects of reading, writing, and mathematics (see the discussion in "Accountability as the Policy Context for Science Education" in Chapter 2). Many schools are aware that they lack adequate science curricula, but do not see this as a high priority relative to other needs.

A comprehensive evaluation of school science curricula by the NSF found that most existing materials did not meet the expectations of the National Research Council's (1996) *National Science Education Standards* (NSF, 1996). They covered too many subjects, included irrelevant classroom activities, and failed to develop important concepts. In contrast, materials considered effective were those that provided students with a sense of purpose about science, engaged them with relevant scientific phenomena, promoted the use of scientific ideas and terms, and encouraged students to examine their own understandings of science. In addition, effective instruction progresses in a sequential manner, with lessons used as building blocks to integrate and expand on developing concepts.

In addition to acknowledging the need for high-quality science curricula for all students, some science educators call for curricula designed for specific student populations, especially those whose cultural beliefs and practices are markedly different from those of Western science or school science. In attempting to make science accessible for all students, the NSF (1998) emphasizes "culturally and gender relevant curriculum materials" that recognize "[diverse] cultural perspectives and contributions so that through example and instruction, the contributions of all groups to science will be understood and valued" (p. 29). However, efforts to develop curriculum materials for culturally and linguistically diverse student groups, as discussed in this chapter, present formidable challenges to science educators. On one hand, developing culturally relevant materials requires a knowledge base from which examples, analogies, and beliefs from a range of different cultures can be drawn and related to specific science topics and scientific practices. Even when such materials are developed and prove effective, their effectiveness may be limited to the particular cultural or linguistic group for which they are designed. On the other hand, materials developed for wide use, particularly computer-based materials that can be accessed through interactive web-based technology, can be implemented across various settings. Yet local adaptations are essential for such materials to be used effectively, which in turn requires expertise on the part of teachers. Thus, there is an inherent tension or trade-off between designing instructional materials that meet the needs of specific local contexts but have limited relevance to other settings, and designing materials that can potentially be implemented across a wide range of settings but require local adaptations.

The small body of literature on science curricula for diverse student groups addresses four areas of research: (a) evaluation of existing curriculum materials (i.e., textbooks or teacher resource manuals) with regard to representation of student diversity, (b) development of culturally relevant materials, (c) use of computer technology in materials development, and (d) use of science curriculum materials with ELL students. Studies in each area reflect a range of research methods, including content analysis of curricular and instructional materials, observations of classroom practices, and experimental designs.

Representation of Student Diversity in Existing Science Curricula

One way to promote science learning and careers for nonmainstream students is to use science curriculum materials and teacher resource manuals to portray scientists from diverse backgrounds, diverse traditions of constructing and transmitting knowledge about the natural world, and information about diverse languages and cultures. Such materials can help render science accessible, relevant, and meaningful for these students. A few studies, described as follows, used content analysis methods to examine representations of human diversity in science curriculum materials for students and teachers.

R. R. Powell and J. Garcia (1985) conducted content analysis of illustrations in seven contemporary elementary science textbook series to examine how they portrayed minorities and females. The results indicate that minority children were represented less frequently[1] than White children (65% White, 31% minority, and 4% unidentified), and minority adults were depicted even less frequently (77% White, 17% minority, and 6% unidentified). Although the textbooks displayed most ethnic groups positively with regard to their involvement in science, adult minorities were usually shown in roles or activities dealing with parental or familial situations and in such occupational roles as teachers and mechanical workers; they appeared less often in science-related career roles.

P. Ninnes (2000) used discourse analysis techniques to examine the approach taken to "minority group knowledges" in two recently published sets of junior secondary science textbooks (one used in Australia and the other in Canada) with a specific focus on the incorporation of indigenous knowledges into the texts. The analysis revealed that each text presented substantial amounts of information about indigenous knowledges, especially related to myths and legends, technology, the natural world, and social activities. The indigenous knowledges reflected a wide range of science topics, such as biology, physics, chemistry, and earth and space sciences. The results also indicate, however, that a number of problems remained unresolved. For example, essentializing indigenous identities through the use of generic and homogenizing labels could inadvertently reproduce racist stereotypes. The representation of particular indigenous lifestyles as "traditional" could be problematic, since a discourse of "traditionality" could be viewed as a means by which non-indigenous people create and control identities for indigenous people. These problems provide insights into the weaknesses of a multicultural science curriculum in particular and multicultural science education broadly. The researcher argues that in addition to greater representation of human diversity in science curriculum materials, more thoughtful and sensitive representations are necessary. The notion that greater representation of diversity in curriculum development is a positive step needs to be problematized by analyzing the kinds of representations that are made. The

---

[1] Minorities are by definition less prevalent, and thus would be expected to appear less frequently in textbooks intended to be representative of the population as a whole. A more relevant statistic would be the ratio of minority portrayals in textbooks to their actual proportion within the general population. In this regard, the numbers provided here concerning portrayals of minority children appear to *over*represent them (probably due to a conscious effort to show diversity). Similarly, the greater prevalence of White adults in the textbooks may reflect the social reality that Whites are overrepresented in science professions, compared to other ethnic groups (with the exception of Asian Americans).

researcher also argues that it is important to emphasize the legitimacy[2] of various ways of knowing the world and the worthwhile contributions of indigenous peoples to this knowing, rather than to simply reproduce the subjugation of indigenous knowledges and the privileging of Western science.

K. Y. Eide and M. W. Heikkinen (1998) conducted content analysis of teachers' editions of 21 middle school science textbooks (i.e., resource manuals) to determine how multicultural information contained therein related to science teaching in multicultural classrooms. The study focused on: (a) the extent of multicultural information in the textbooks, using 75 multicultural descriptors (e.g., ethnicity, culture, multiculturalism, diversity, minority, bilingual, cross-cultural); (b) the distribution of multicultural information within 10 foundational knowledge categories (philosophy, education, religion, economics, medicine, history, sociology, psychology, vocations, and politics); and (c) the relationship of multicultural information to science content. The results indicate that support for multicultural information varied remarkably among the 21 resource manuals in the study. In most resource manuals, the multicultural information was found in very small amounts, usually two to four sentences that covered approximately one-tenth of a page, inserted at the rate of one per chapter or unit. Additionally, the results indicate that education comprised the largest foundational knowledge category (45%), followed by economics (17%) and history (12%). Finally, the relationship of the multicultural information to the science content revealed that 11.7% was highly related, 49.2% was somewhat related, and 39.0% was not related. The reason for the inclusion of the multicultural information was often not identified, and only in rare cases were suggestions given on how the teacher might use or structure the information to support the learning objectives indicated in the teachers' manual.

The studies just described indicate that the portrayal of minorities in science-related career roles in textbooks is limited (Powell & Garcia, 1985), that problems remain unresolved even in materials designed to emphasize indigenous knowledges (Ninnes, 2000), and that cultural diversity is not adequately represented in teacher resource manuals (Eide & Heikkinen, 1998). Considering the importance of role models from one's own race/ethnicity and gender in choosing a science career (Grandy, 1998; Hill et al., 1990), the limited representation of cultural diversity in most science curriculum materials in the United States raises concerns.

[2] Note that this legitimacy may be based on criteria other than scientific rigor, e.g., strengthening of cultural identity or societal bonds, or adherence to an ecologically sustainable way of life (even if this adherence is not based on accurate understandings of scientific phenomena). See the earlier section on "Views of Science" in Chapter 2.

Culturally Relevant Science Curricula

Faced with the dearth of science curriculum materials designed to be culturally relevant to nonmainstream students, a small number of science education researchers have developed materials incorporating experiences, examples, analogies, and values from specific cultural and linguistic minority groups. Two such efforts are described in this section.[3]

G. S. Aikenhead (1997, 2001b) offers a conceptual framework for designing culturally relevant curriculum materials, based on the notion of "cultural border crossing" between students' everyday world and the culture of science. On the basis of this framework, Aikenhead describes the development for grades 6–11 of curriculum units that integrate Western science with "Aboriginal sciences" (Aikenhead's term) of First Nations groups in northern Saskatchewan, Canada. The units identified two important cultural contexts: that of students' Aboriginal communities and that of Western science and technology. Throughout the units, both Western scientific and Aboriginal values were made explicit. Each lesson plan identified a scientific value (e.g., control over nature) and/or an Aboriginal value (e.g., harmony with nature). The units gave students access to Western science and technology without requiring them to adopt the worldview endemic to Western science or change their own cultural identity. Based on a bicultural and bilingual model, the units encouraged students to traverse cultural borders between the realm of Western science and their own cultural identity. Informal assessment of classroom practices indicated that students participated in these units in ways that were culturally meaningful to them (but no specific information about assessment results is provided).

In response to the low science achievement of Native American students, as measured by standardized tests, C. E. Matthews and W. S. Smith (1994) tested the effect of culturally relevant materials (see their article for details about these materials) on the achievement and attitudes of Native American students in grades 4 through 8 at Bureau of Indian Affairs (BIA) schools. Although a vigorous attempt was made to produce a representative sample of the 103 BIA-operated schools, this was not achieved because some schools declined to participate in the study and some did not return complete or usable data. The final sample included nine schools from eight of the BIA's agencies. The students in the sample were 60% Navajo, 17% Sioux, 9% Papago, 7% Hopi, and 7% from other tribes. Within a pretest-posttest control group design, the study tested the effect of the intervention on two student outcome variables: science achievement and attitudes toward

---

[3] Lipka and Adams (2004) developed a culturally based mathematics curriculum as a way to improve Alaska Native students' mathematics performance. Using a quasi-experimental design, the study demonstrated a statistically significant impact of the curriculum on Yup'ik students' understanding of the mathematical topics under study, in one urban and four rural school districts.

Native Americans and school science. Teachers were randomly assigned to the experimental or control group by predetermined procedures. The intervention was carried out over a 10-week period, during which the teachers who used the culturally relevant materials were to teach science for 25 hours and related language arts (i.e., profiles of Native Americans who use science in their daily lives) for 25 hours. Teachers in the control group were to teach science, using the same instructional materials as the other group but without the Native American references, for 25 hours and their usual language arts for 25 hours. However, teacher logs and telephone conversations revealed that instead of the intended 50 total hours during the 10-week intervention period, the experimental group teachers used the culturally relevant materials an average of 33 total hours, and the control group teachers used the other materials, on average, less than 25 total hours, and less than 10 hours in most cases. Multivariate analysis of covariance (MANCOVA) tests were conducted for three independent variables (treatment, tribe, and gender), using the pretest scores as the covariates. The results indicated that students who were taught science via the culturally relevant materials showed significantly higher achievement scores and displayed significantly more positive attitudes toward both Native Americans and science than did comparable students who were taught science without the culturally relevant materials. However, the results should be taken with caution, considering that the experimental group spent more hours than the control group.

Technology-Based Science Curricula

In addition to the text-based curriculum materials described, several studies developed computer-based curriculum materials and examined their impact on students' science outcomes. These studies, described below, were conducted by two research teams that have engaged in programmatic lines of research over the years. Both research projects targeted middle school students in urban school districts. In contrast to culturally relevant materials designed for specific groups, these computer-based materials, accompanied by interactive web-based technology, are intended for large-scale implementation, although local adaptations are necessary for effective use. The results show positive changes in student achievement on standardized tests.

The "Learning Technologies in Urban Schools" [LeTUS] project is aimed at developing curriculum materials that would be usable in urban settings and scalable across entire school districts. In collaborative partnership with teachers in Detroit and Chicago, the research teams at the University of Michigan and Northwestern University developed project-based curriculum materials (grounded in a social constructivist perspective) that contextualize science learning in meaningful and real-world problems, engage students in science inquiry, and use computer technology to support science inquiry. Over the years, the project has developed a series of science units for middle school students and examined the impact of science curriculum materials accompanied by learning technologies on science achievement in urban school districts. These units were designed by collaborative teams and were revised yearly on the basis of student outcomes and teachers' experiences in implementation. The units were aligned with national and district curriculum frameworks and serve the district's urban systemic reform program in science. Consistency of implementation across the participating schools was limited by the fact that not all the schools had access to the same level of computer technology for classroom use, reliability and capacity of machines varied considerably, maintenance was not always timely, and internet access was unreliable. In addition to the curriculum units and learning technologies, the project offered professional development opportunities for teachers. In most schools, one to three science teachers participated, based on their interest or because they were selected by their school administration.

R. W. Mark and colleagues (2004) reported student achievement results with nearly 8,000 middle school students (grades 6 through 8) from 14 schools representing the broad range of schools and neighborhoods in Detroit over three years of the LeTUS project. The data consisted of pretest–posttest gain scores based on project-developed achievement measures for four curriculum units (one for 6th grade, two for 7th grade, and one for 8th grade). The researchers examined whether student outcomes improved as the project was implemented with larger numbers of teachers and in larger numbers of classrooms over the years (i.e., scaling-up). The results showed statistically significant increases on curriculum-based test scores for each year of teacher participation. Furthermore, the impact of the innovation continued to grow while scaling-up occurred, as evidenced by increasing effect size estimates over the years. The results indicate that students who are historically low achievers in science can succeed when provided with inquiry-based and technology-infused curriculum units along with professional development of teachers.

As part of the LeTUS project, A. E. Rivet and J. S. Krajcik (2004) focused on the 6th-grade unit about machines, which was designed to relate science to African American students' experiences in their community. This unit was set in the context of developing a new machine to construct large buildings and bridges. After discussing large structures in the city, students took a walking tour of an active construction site near the school and described the different machines they saw and how these machines functioned to help people build large buildings. Students used this anchoring experience to develop a design for a new machine of their own invention. Over the three-year period of the project, 2 teachers participated during the first year, 4 teachers during the second year, and 11 teachers during the third year (some of the teachers participated for more than one year). Students were then assessed via science tests including multiple-choice and short answer items. Dependent (or matched) t-test analyses were conducted to compare the pretest and posttest results after each year of the project. The results showed statistically significant and consistently high achievement gains in both understanding of science concepts and inquiry process skills.

The "Kids as Global Scientists" Weather program by N. B. Songer (H.-S. Lee & Songer, 2003; Songer, H.-S. Lee, & Kam, 2002; Songer, H.-S. Lee, & McDonald, 2003) involves a technology-based middle school science curriculum along with teacher professional development. Using mixed methods, these studies examined the impact of an inquiry-based, technology-rich middle school learning environment focused on weather in a large, predominantly African American urban school district, as well as other participating schools across the nation. Originally implemented with a limited number of self-starter teachers and well-supported school contexts, the curriculum was implemented simultaneously with approximately 230 teachers (some of whom taught multiple classes) and 13,000 students in 4th through 9th grade from 40 states.

The study by Songer and colleagues (2003) addressed scaling-up of this learning environment enacted simultaneously in hundreds of classrooms across the nation. It involved two groups of teachers. One group consisted of 40 "maverick" teachers distributed throughout the nation, who sought out the program, customized it according to their own needs, and did not receive systematic professional development. These teachers tended to work in schools with a relatively rich fund of resources and supports. The other group consisted of 17 teachers from a recent partnership between the project and a large, high-poverty urban school district. This group of teachers received targeted professional development to address obstacles common among their respective schools. From the total sample of 57 classrooms, five "successful" classrooms were selected for detailed analysis, on the basis of achievement gains from pre- to posttests using multiple-choice and open-ended items. These successful classrooms included three in high-poverty urban environments and two "maverick" classrooms in middle-class suburban environments. Dependent t-tests were conducted with pre- and posttest scores of students in each classroom. After completion of the eight-week inquiry-based weather program, students in both settings demonstrated achievement gains in scientific inquiry and content knowledge measured on both multiple-choice and open-ended items.

Although the program itself and patterns of student achievement were similar among the five classrooms, the classroom practices that led to these outcomes were somewhat different in each case. Self-reports of all 57 teachers at the beginning and end of the eight-week program indicated two distinguishing features: (a) classroom use of small or large groups and (b) partial or full inquiry. Almost half of the "maverick" teachers favored having students work in small, self-paced groups with relative auton-omy, whereas most urban teachers usually had the whole class doing the same activity in unison and tended toward a more teacher-directed peda-gogy. The researchers interpreted the differences between the two groups in terms of class size, students' prior experience with science inquiry, and institutional resources and support structures. Rather than calling for student-initiated science inquiry as a pedagogical ideal (as science stan-dards documents tend to do), they claim that different versions of science inquiry or instruction should be adapted for different types of classrooms.

Science Curricula for ELL Students

The literature on science curriculum materials for ELL students addresses the same three issues discussed earlier: (a) evaluation of existing curricu-lum materials (i.e., textbooks or teacher materials), (b) development of culturally and linguistically relevant materials, and (c) use of computer technology in materials development.

On the basis of observations of 57 randomly selected elementary bilingual/bicultural classrooms serving predominantly Hispanic/Latino students in a large metropolitan area of the southwestern United States, R. H. Barba (1993) reported that the students received science instruc-tion using materials that were not relevant to their language and culture. No classrooms had Spanish-language textbooks available for student use. Although 61% of the classrooms had science kits available in Spanish, English, or both languages (including the bilingual science kit, the *Finding out/Descubrimiento* materials, by DeAvila, Duncan, & Navarrete, 1987a, 1987b), these materials were used for instructional purposes in only 6 (10.5%) of the 57 classrooms. In these 6 classrooms, manipulative materials were used only 12% of the instructional time.

Several studies examined the impact of various curriculum materials on ELL students. S. Lynch, and collegues (2005) examined the effects of a highly rated science curriculum unit on a diverse student popula-tion. The curriculum unit, which used a traditional textbook format, was not designed for the purpose of cultural or linguistic relevance to spe-cific groups; instead, it was designed for wide implementation. The unit was implemented among more than 1,000 8th-grade students in five mid-dle schools selected for racial/ethnic, linguistic, and economic diversity, whereas a variety of district-approved curriculum materials were used with 1,200 students in the five middle schools that formed the comparison group. The quasi experiment found statistically significant achievement gains in students' science outcomes on a standardized test. Disaggregated achievement data indicated that subgroups of students in the treatment condition outscored their comparison group peers in all cases, except for students currently enrolled in ESOL.

E. Hampton and R. Rodriguez (2001) implemented a hands-on, inquiry science curriculum (i.e., the Full Option Science Series, FOSS) with Spanish-speaking elementary children who were developing English fluency along with their first-language skills. Using this curriculum, university interns taught science to students in kindergarten through 5th grade in 62 class-rooms at three elementary schools near the U.S.-Mexican border. They taught six one-hour lessons over the course of six weeks, with half of the instruction in Spanish and half in English. One written assessment, con-taining three inquiry items and three open-ended response items about the Foods and Nutrition unit, was administered to 107 5th-grade students. The four-page written assessment was available to the students in Spanish or English, and they could respond in the language of their choice. Of the students, 55% chose to respond in Spanish and 45% responded in English. Correct performance ranged from about 33% to 51% across the six items. There was relatively little difference between children who chose to respond in Spanish and those who chose to respond in English. Addition-ally, participants' perceptions were examined from multiple data sources, including university interns via written comments and focus group inter-views, in-service teachers via an attitude survey and written comments, and 80 3rd-grade students via an attitude survey. The consistency of the data indicates that there was a strong positive feeling among university interns, classroom teachers, and elementary students about the value of this inquiry approach for increasing the children's understanding of sci-ence concepts in both languages.

As part of their ongoing research, O. Lee and colleagues devoted con-siderable effort to developing curriculum materials in order to imple-ment a professional development intervention with teachers, which subse-quently translated into classroom practices with students. Over the years, the project developed curriculum materials for 3rd-, 4th-, and 5th-grade students. These included units on measurement and matter for 3rd grade, the water cycle and weather for 4th grade, and the ecosystem and solar system for 5th grade. These topics follow the sequence of instruction from basic skills and concepts (measurement and matter), to variable global sys-tems (the water cycle and weather), to increasingly large-scale systems (the ecosystem and the solar system).[4] The curriculum materials for each sci-ence topic include science booklets for students, teachers' guides (including

---

[4] In their current research, Lee and colleagues are expanding curriculum development efforts to develop comprehensive science curriculum units for grades 3 through 5 students and accompanying teachers' guides. The materials are designed to promote ELL students' sci-entific reasoning and inquiry, while also preparing them for statewide science assessments (in English).

transparencies), and class sets of supplies (including trade books related to the science topics in the units). All the units emphasize three domains: (a) science inquiry, progressing along a continuum from teacher-explicit instruction to student-initiated inquiry (for details, see Lee, Hart, Cuevas, & Enders, 2004); (b) integration of English language and literacy development in science instruction (for details, see Hart & Lee, 2003); and (c) incorporation of students' home language and cultural experiences in science instruction (for details, see Luykx et al., 2005).

S. H. Fradd and colleagues (2002) describe the development, implementation, and impact of curriculum units on matter (culminating in the water cycle) and weather. The units were implemented with more than 500 4th-grade students from different ethnolinguistic backgrounds (Hispanic, Haitian Creole, and monolingual English-speaking students of White and African American descent) at six elementary schools in a large urban school district in a southeastern state. Instruction of each unit took approximately eight weeks, and most teachers taught the matter unit in fall and the weather unit in spring. At the beginning and end of each unit, students completed a paper-and-pencil test containing multiple-choice, short-answer, and extended written response items. Dependent t-test analysis of pre- and posttest scores indicates that students from all ethnolinguistic groups showed statistically significant achievement gains in science knowledge and inquiry, respectively.

A few researchers examined the use of interactive, computer-based curriculum materials with ELL students. C. Buxton (1999) used student-generated computer models as a medium for elementary students to develop meaningful explanations of science content. The study was based on a qualitative analysis of students' engagement in computer modeling in a two-way bilingual classroom. This combined 2nd- and 3rd-grade classroom consisted of roughly half native-English speakers and half native-Spanish speakers. The results indicate that even for primary grade students with limited prior exposure to computers, the use of student-generated computer models in conjunction with the construction of physical models and other hands-on activities provided meaningful opportunities for students to think, act, and talk in science. The researcher pointed out factors that aided in students' ability to engage in scientific discourse: (a) classroom discussions and activities aimed at helping students connect science to their own personal experiences, (b) greater depth of coverage of science topics, (c) the creation of settings that encouraged students to talk about the science content demonstrated in their computer models, and (d) settings that allowed students to code-switch in the context of scientific discourse. However, students' ability to engage in scientific discourse was hindered by such factors as frustration arising from the difficulty of mastering the nuances of the computer software, and the frequent need to focus on the "how to's" of model construction rather than on the science content being modeled. Despite these difficulties involved in learning how to use technology, experiences of this kind are especially valuable for ELL students, who have historically been excluded from meaningful science learning experiences.

J. K. Dixon (1995) employed a quasi-experimental research design involving four 8th-grade classes in the treatment group and five classes in the control group. The three independent variables were the students' level of English proficiency, their level of experience with computers, and their level of visualization ability. The study used four posttests as outcome measures, including two content measures (a paper version and a computer version) of the concepts of reflection and rotation and two measures of visualization ability. The two measures of students' visualization ability along with a language assessment battery also served as three covariates. The intervention and data collection were conducted as follows. First, at the beginning of the study (prior to data collection), all classes in the treatment and control groups were taught how to use the computer software program under investigation. Second, tests of the three covariates were administered to all nine classes. Third, students were instructed using the reflection and the rotation units for approximately two weeks. Students in the treatment group spent that time learning how to use the computer software program in order to conjecture about and construct knowledge of reflections and rotations. Students in the control group were presented with content on reflection and rotation using the traditional, teacher-directed, textbook approach. Finally, at the completion of the unit, the four outcome measures were administered to each group. (The researcher's article does not provide any information about the language treatment of the two groups, with regard to either instruction or the administered outcome measures. However, the article does mention that student pairs consisting of Spanish speakers conversed in their home language and that students were allowed to demonstrate their understanding in Spanish or English while working on the computer program.) An analysis of covariance (ANCOVA) was used to control for initial differences between the two groups. After controlling for initial differences, students experiencing the dynamic, computer-based instructional environment significantly outperformed the control group on all four outcome measures. In both the treatment group and the control group, there was no statistically significant difference between ELL students and English-proficient students on any of the four outcome measures when they experienced the same instructional environment.

Discussion

Research on science curricula for nonmainstream students addresses a number of questions using a range of research methods. However, the literature is very limited. In efforts to develop culturally relevant science curricula, Aikenhead (1997, 2001b) and Matthews and Smith (1994) present detailed descriptions about the curriculum development process and how they articulated science content and process with the cultural beliefs and practices of specific groups. Given that the studies either did not provide information about science assessment (Aikenhead, 1997, 2001b) or suffered limitations in carrying out an experimental research design (Matthews & Smith, 1994), it is difficult to conclude whether the curriculum materials employed indeed positively impacted student outcomes in terms of higher achievement in science, more positive attitudes toward science, or enhanced cultural identity among nonmainstream students.

The "LeTUS" project and the "Kids as Global Scientists" Weather project highlight challenges in large-scale implementation of technology-based science curricula. Since the LeTUS project was designed to involve all the middle schools in a school district, it was not feasible to conduct an experimental study (Mark et al., 2004; Rivet & Krajcik, 2004). Even if an experimental study could be designed, it would present the ethical dilemma of which schools would benefit from the intervention (i.e., the experimental group) and which schools would be left out of the intervention (i.e., the control or comparison group). Even supposing that these ethical problems could be resolved, high student mobility and teacher attrition presents formidable challenges to carrying out research. Such issues and concerns become more complicated when a technology-based science curriculum is implemented widely across the nation, as Songer and colleagues have done (H.-S. Lee & Songer, 2003; Songer et al., 2002; Songer et al., 2003). On the one hand, the program is striving toward its goal of large-scale implementation nationwide. On the other hand, this entails risks to the researchers' control of the research design, systematic data collection, and the capacity to handle large data sets. Conducting such research requires extensive funding, which is scarcer for science education than for core subjects of reading, writing, and mathematics. Given these limitations, results from these two technology-based projects must be interpreted with caution. However, the rigor and evidentiary warrants must be evaluated in the context of the constraints inherent in large-scale research. Future research efforts may focus on how to uphold the rigor of research within such constraints.

While a culturally relevant science curriculum focuses on developing materials for specific student groups, standardized science curriculum is intended for large-scale implementation across a wide range of student groups or educational settings. The goal of localization using a culturally relevant curriculum and the goal of large-scale implementation using a standardized curriculum each present unique sets of challenges. The demand for localized knowledge in a culturally relevant curriculum reduces its applicability to student groups other than those originally intended, whereas large-scale implementation of a standardized curriculum requires adaptations and modifications for different educational settings.

To ameliorate the lack of culturally relevant materials for ELL students, efforts are being made to develop science curriculum materials for these students. Hampton and Rodriguez (2001) tested the impact of a commercially available FOSS science curriculum that fosters hands-on, inquiry science. Fradd and colleagues (2002) developed and tested materials that integrate science inquiry, students' home language and culture, and English language and literacy development. Buxton (1999) and Dixon (1995) designed and tested computer-based curriculum materials for ELL students. Through these interventions, ELL students learned to engage in scientific discourse (Buxton, 1999), made positive achievement gains in both science knowledge and inquiry (Fradd et al., 2002), performed comparably when they chose to respond either in English or in their home language (Hampton & Rodriguez, 2001), and performed comparably to English-proficient students (Dixon, 1995). Although the results are promising, caution is warranted in drawing conclusions based on this limited literature.

5

Science Instruction

All students come to school with knowledge constructed within their home and community environments, including their home language(s) as well as cultural beliefs and practices. Learning is enhanced – indeed, made possible – when it occurs in contexts that are culturally, linguistically, and cognitively meaningful and relevant to students. Effective science instruction must consider students' home cultures and languages in relation to the pedagogical aims of science instruction. Reviews of literature on effective instruction have focused on nonmainstream groups in general (Garaway, 1994; Lee, 2002, 2003; Lee & Fradd, 1998) as well as specific groups, including African American (Atwater, 2000), Asian American (Lee, 1996), Hispanic (Rakow & Bermudez, 1993), and Native American students (Nelson-Barber & Estrin, 1995, 1996; Riggs, 2005).

Beyond the literature on science learning, described in Chapter 3, there is a rather extensive body of literature on science instruction with nonmainstream students. Since learning and instruction are closely related, these two areas of literature are guided by common theoretical/conceptual perspectives (cross-cultural, sociopolitical, cognitive science, etc.). Within each perspective, some studies examine existing instructional practices, whereas others report on the design and implementation of instructional interventions and their impact on teachers and/or students. Although intervention efforts generally emphasize the articulation between science disciplines and some aspect of student diversity, how this articulation is carried out differs, depending on the specific points of contact and/or conflict between students' home cultures and the culture of science.

Culturally Congruent Science Instruction

Children from nonmainstream backgrounds acquire cultural norms and practices in their homes and communities that are sometimes incongruent with those of school. Teachers therefore need to be aware of a variety of cultural experiences in order to understand how different students may approach science learning. Teachers also need to use cultural artifacts, examples, analogies, and community resources that are familiar to students in order to make science relevant and intelligible to them.

Numerous studies suggest that when students receive culturally congruent instruction, they respond positively in terms of improved verbal and academic performance (e.g., Au, 1980; Deyhle & Swisher, 1997; Heath, 1983; Tharp & Gallimore, 1988). The literature on cultural congruence has traditionally focused on classroom interaction, communication, and literacy development (Gay, 2002; Ladson-Billings, 1994, 1995; Osborne, 1996; Villegas & Lucas, 2002). Recently, a few studies have focused on subject areas, including literature (e.g., C. D. Lee, 2001), mathematics (e.g., Brenner, 1998; Lubienski, 2003), social studies (e.g., McCarty et al., 1991), and science (e.g., Lee & Fradd, 1998; Warren et al., 2001). The following discussion addresses science instruction in relation to the cultural experiences and practices of nonmainstream students.

Incongruent Instruction
A. Contreras and O. Lee (1990) examined how middle school science teachers showed differential treatment to students of different cultural backgrounds, and how these practices influenced students' development of science knowledge and skills both within the classroom and in relation to outdoor science activities. Using ethnographic methods, the research followed two middle school science teachers (one White male and one African American female), each of whom taught an enriched class and a regular class over the course of the school year. The students in the White male teacher's enriched class were predominantly White, whereas about half of the students in his regular class were African American or Hispanic. The majority of the students in both of the African American teacher's classes were White. The White male teacher treated his two classes differently, often failing to provide nonmainstream students in his regular class with meaningful classroom activities and indirectly preventing them from participating in science field trips (due to school policies prohibiting participation of students with poor citizenship grades). Thus, classroom practices and school policies seemed to exacerbate the "cultural gap" that already existed between the teacher and his students and among students of different backgrounds. The African American female teacher, in contrast, explicitly stated her efforts to be fair to all students and treated students in both classes similarly in terms of classroom activities and citizenship grades.

C. Westby and colleagues (1999) describe communication and interaction patterns between Hispanic and Haitian American elementary teachers and their students in classrooms where teacher and students shared a similar linguistic and cultural background.[1] The study involved one Haitian American and three Hispanic teachers and their 4th-grade students at two inner-city schools in a large urban school district. As part of a larger research project, the four teachers all taught the same instructional unit, which was developed by the project. This study was based on microanalysis of videotapes of a lesson on a water cycle simulation activity, with regard to (a) quantitative features of teachers' and students' utterances and (b) science learning components in terms of knowing, doing, and talking science. The teachers and students engaged in culturally congruent interaction patterns during science classes. For example, the three Hispanic teachers used social talk to relate personal experiences to the academic content, communicated a sense of concern for the well-being of the children, and made humorous comments that appeared to create a positive learning atmosphere conducive to student participation. The Haitian American students were much less familiar with working collaboratively in small groups and expected more direct and explicit guidance from the teacher. Across all four classrooms, both the Hispanic and Haitian American students were gaining skills in "knowing science" and "doing science." However, they needed to master "talking science," that is, using the academic, descriptive, explanatory, and argumentative genres of scientific discourse. The results suggest that in addition to establishing culturally congruent interaction patterns, teachers need sufficient knowledge of science to teach effectively. These results, while seemingly obvious, highlight the limitation of the existing literature, which often addresses classroom participants' cultural patterns and disciplinary knowledge separately, rather than examining the intersection of the two.

E. Moje and colleagues (2001) describe a bilingual middle school science teacher with predominantly Spanish-speaking students in an urban school in a large school district. As part of a larger research project, the teacher taught a project-based science curriculum to promote students' scientific inquiry. Although the school administration expressed a commitment to two-way bilingual education, all instruction was conducted in English, under the justification that none of the area high schools offered bilingual programs. The researchers conducted discourse analysis of science instruction over the course of the school year. Although the teacher had extensive science knowledge and his linguistic and cultural background was similar to that of his students, he often had difficulties articulating students' everyday knowledge and primary discourse with scientific knowledge and discourse. The results suggest that in order to assist students in constructing

---

[1] This "shared background" is much more broadly defined in the case of the Spanish speakers, whose nationalities and other circumstances were more diverse than those of the Haitian American participants.

new knowledge, teachers need to establish spaces in which different discourses and knowledges – from science disciplines, the science classroom, and students' lives – are brought together.

C. Buxton (2005) described one such attempt to bring together competing discourses of school, science, and students at an urban science and mathematics magnet high school. He looked at how teachers gradually changed their expectations of what their students believed to be high-quality academic work, as they made explicit for their students what quality academic work entailed (e.g., organized, clear and complete, evidence-based, turned in on time, individually responsible, good or poor responses). He also looked at how the teachers redefined their roles and responsibilities with regard to helping their students engage in quality academic work. This three-year ethnographic study used a conceptual framework grounded in cultural models of identity formation, positing student and school cultures as fluid rather than static models for interpreting the world. Analysis of field notes, interviews, and classroom artifacts indicated an evolving culture of academic success and a similarly evolving model of an "educated person." Teachers and students collaboratively constructed and negotiated this model based on their interpretations of learning, achievement (short-term), resistance, and academic success (long-term). He found that teachers and students who remained at the magnet school for more than two years came to share a largely overlapping discourse of academic success based on the model of an "educated person," whereas teachers and students who left the school within two years of arriving retained models of academic success that were largely incompatible with the model shared by those who remained. He recommends that teachers examine both competing and overlapping discourses originating in school, science, and students, and find ways to enhance the cultural congruence in their teaching.

Congruent Instruction

School science assumes certain prior knowledge on the part of students, not only of science content and process but also of the types of interaction and discourse through which science learning is believed to occur. In science classrooms, students are expected to ask questions, carry out investigations, find answers on their own, and formulate explanations in scientific terms. These practices are essential to scientific inquiry, but are not equally encouraged in all cultures (Arellano et al., 2001; Atwater, 1994; Jegede & Okebukola, 1992; McKinley et al., 1992; Solano-Flores & Nelson-Barber, 2001). Additionally, the discourse patterns and verbal and written registers associated with scientific inquiry may be less familiar to some students (and some teachers) than to others (Lemke, 1990; Moje et al., 2001).

In a programmatic line of research conducted since the early 1990s, O. Lee and colleagues have extended the notions of cultural congruence and culturally relevant pedagogy to propose the framework of "instructional congruence," with the aim of articulating science disciplines with students' languages and cultures (for conceptual discussion, see Lee, 2002, 2003; Lee & Fradd, 1998; for methodological discussion, see Luykx & Lee, in press). This framework highlights the importance of developing *congruence*, not only between students' cultural expectations and norms of classroom interaction but also between students' linguistic and cultural experiences and the specific demands of particular academic disciplines, such as science. The need to articulate these two domains is especially critical when they contain potentially discontinuous elements. Thus, instructional congruence emphasizes the role of *instruction*, as teachers (or educational interventions) explore the relationship between academic disciplines and students' cultural and linguistic knowledge and devise ways to link the two. This framework can serve as a conceptual and practical guideline for curriculum materials development, teacher professional development, classroom practices, teacher change, and student assessment.

Students' cultural beliefs and practices are sometimes discontinuous with Western science; therefore, effective science instruction should enable students to cross cultural borders between their home cultures and the culture of science (Gao & Watkins, 2002; George, 1992; Jegede & Aikenhead, 1999; Jegede & Okebukola, 1991a; Loving, 1998; Shumba, 1999; Snively, 1990). According to the multicultural education literature, school knowledge represents the "culture of power" of the dominant society (Au, 1998; Banks, 1993a, 1993b; Delpit, 1988, 1995; Reyes, 1992). The rules of classroom discourse are largely implicit and tacit, making it difficult for students who have not learned the rules at home to figure them out on their own. For students who are not from the culture of power, teachers need to provide explicit instruction about that culture's rules and norms for classroom behavior and academic achievement. Without this explicit instruction, the students lack opportunities to acquire the rules, as well as the access to learning opportunities.

As students gradually acquire the cultural competencies needed for academic achievement, they may require explicit instruction of academic content if they are to obtain the high-status knowledge that their more privileged peers have access to outside the classroom. Explicit instruction of content in the context of authentic and meaningful tasks and activities has been advocated with nonmainstream students in literacy instruction (e.g., Au, 1998; Delpit, 1988; Jiménez & Gersten, 1999; Reyes, 1992); literature instruction (e.g., C. D. Lee, 2001); mathematics instruction (e.g., Brenner, 1998; Lubienski, 2003); and science instruction (e.g., Fradd & Lee, 1999; Lee, 2003).

Explicit instruction seems to imply at least two notions (Delpit, 1988; Fradd & Lee, 1999; Lee, 2002, 2003). First, it requires instructional scaffolding to make explicit the transition from one set of values and practices to another. Teachers need to make visible students' everyday knowledge, the relationship between students' knowledge and academic tasks, and the transition from one domain to the other. For example, teachers may point out for students that questioning and argumentation with teachers and peers is encouraged in the science classroom, although it may not be acceptable with adults at home. The aim is to encourage students to question and inquire without devaluing the norms and practices of their homes and communities, so that students gradually learn to cross cultural borders between the norms and practices of their home and community environments and those of Western science and schooling more broadly.

Second, explicit instruction implies teacher-directed instruction in which teachers tell students what to do or provide extensive guidance as students work on academic tasks. In promoting science inquiry, student-initiated learning, in which students ask questions and find answers on their own, is commonly the instructional goal. The issue is where to start and what to do to reach this goal with students from diverse backgrounds and levels of science experience. For those students who have limited science experience, teachers may need to provide direct instruction to build necessary concepts and skills within the context of meaningful and authentic tasks (Duran et al., 1998; Moje et al., 2001; Songer et al., 2003).

As a means of reaching this goal, Lee and colleagues proposed the *teacher-explicit to student-exploratory continuum*, which takes into account students' cultural backgrounds as well as previous science experiences (Fradd & O. Lee, 1999; Lee, 2002, 2003). Teachers move progressively from more explicit to more student-centered instruction, gradually reducing assistance while encouraging students to take the initiative, explore on their own, and assume responsibility for their own learning. Along the continuum, teachers should consciously maintain a balance between teacher guidance and student initiative, as they make decisions about when and how to foster students' responsibility for their science learning.

Grounded on the notions of instructional congruence and the teacher-explicit to student-exploratory continuum, Lee (2004) examined how elementary teachers mediated school science with the linguistic and cultural experiences of their students. The research involved six Hispanic elementary teachers (all fluent in English and Spanish) who taught Hispanic students from various racial and national backgrounds in a large urban school district. The teachers were recommended by their principals for excellence in teaching and commitment to their students. During three years of collaboration, the teachers participated in professional development opportunities and the design of instructional units (see Fradd et al., 2002). Data collection and analysis of classroom observations, interviews, and questionnaires focused on teachers' beliefs and practices with regard to science instruction, incorporation of students' language and culture in science instruction, and English language and literacy development as part of science instruction. The teachers realized that while students' linguistic and cultural practices constituted potential intellectual resources for science learning, such practices sometimes conflicted with scientific practices. They noted that some cultures might not promote questioning or exploration by children. Teachers also found it difficult to relinquish their own authority and control in favor of enhancing student autonomy. Additionally, they pointed out the conflict between their students' preference for group collaboration and the need for independent performance. Over the course of the school year, the teachers initially emphasized explicit instruction, whole-group participation, and teacher authority and control. In guiding students through explicit instruction, the teachers orchestrated the class as a whole. Even when students worked in small groups, the teachers organized the groups as part of the entire class and encouraged collaboration and teamwork. Gradually, they enabled students to take the initiative in conducting science inquiry, promoted student autonomy, and encouraged students to work individually and independently while also valuing the teamwork and collaboration that most students preferred.

Cognitively Based Science Instruction

The cognitive science perspective sees the relationship between scientific practices and students' sense making in a complex and reflexive way – as similar, different, interactive, and generative (Brown, 1992, 1994; diSessa et al., 1991; Lehrer & Schauble, 2000). The entry point for effective teaching is to examine the everyday experiences and informal language practices that individual students bring to the learning process. Students have developed forms of reasoning and argumentation in their everyday lives that can serve as intellectual resources in academic learning. A major problem in teaching is that teachers fail to recognize the diverse ways in which these intellectual resources manifest themselves. When their own intellectual and cultural resources are marginalized from the learning process, students may withdraw from that process and have fewer and fewer opportunities to learn in school. This problem occurs more often with nonmainstream than with their mainstream counterparts because of differences between students' and teachers' cultural practices and classroom expectations (Warren et al., 2001).

Once individual students' everyday experience and informal language are recognized as cultural and linguistic resources relevant to science tasks, the intersections between students' everyday knowledge and scientific practices can be examined (Ballenger, 1997; Warren & Rosebery, 1995, 1996; Warren et al., 2001). Teachers need to relate, enlarge, and elaborate on these intersections or areas of contact between what students know and know how to do, on the one hand, and scientific knowledge and practices, on the other. Only then can the potential of students' own intellectual resources be harnessed to the purposes of science instruction.

The Chèche Konnen Project promotes collaborative scientific inquiry among language minority and low-SES students (Rosebery et al., 1992; Warren & Rosebery, 1995, 1996). The premise is that much can be learned about school science by examining science as it is practiced in professional communities. Although scientific practice in schools may not – and perhaps should not – mirror the scientific practice of actual research scientists, understanding the relationship between these two domains can help clarify what it means to teach and learn science. To promote science inquiry and argumentation, teachers involved in the project provide students with opportunities to engage in collaborative scientific inquiry. On the basis of a model of what scientists do in the real world (although in a much simplified form), students learn to use language, think, and act as members of a science learning community.

In the Chèche Konnen Project, the course of students' inquiry is not predetermined; rather, it grows directly out of students' own beliefs, observations, and questions. The investigation of one question leads to additional explorations initially unforeseen. Because science instruction is organized around students' own observations and interests, the "curriculum" emerges from the questions students pose, the experiments they design, the arguments they engage in, and the theories they articulate. The teachers' role is to facilitate students' investigations of their own questions, while offering guidance and assistance as needed.

The results of the project indicate that students with limited English proficiency or limited science experience were capable of conducting scientific inquiry and appropriating scientific ways of knowing and reasoning after participating in science instruction designed to promote collaborative scientific inquiry. For example, Rosebery et al. (1992) examined the effects of "doing science" on language minority students' appropriation of scientific ways of knowing and reasoning. The study involved students from a variety of linguistic and cultural backgrounds in one middle school and one high school classroom. The study was based on interviews with 16 students at the beginning and end of instruction, and student responses were analyzed using qualitative and quantitative (paired t-tests) methods. In September, students showed almost no evidence that they understood what it means to reason scientifically and, specifically, to put forward hypotheses having deductive consequences that can be evaluated through experimentation. Students used personal experience as evidence for a particular belief, rather than using the discourse of conjecture and experimentation that calls for critical, analytic evaluation of given information or evidence. Throughout the school year, the students engaged in scientific inquiry, such as analyzing the water quality of fountains on a school ground or the ecology of a local pond. By June, the students were able to go beyond the information given to put forward hypotheses that were explanatory and testable. They were aware that hypotheses drive scientific inquiry and that experimentation is a means for developing evidence.

Over the years, while expanding the view of science as reflexive and cognitively complex, research by the Chèche Konnen team has considered the informal, everyday knowledge that students of diverse backgrounds bring to the learning process (Ballenger, 1997; Warren et al., 2001). Teachers need to understand the complex dynamics between scientific practices and students' everyday knowledge. Students from many different languages and cultures deploy sense-making practices – deep questions, vigorous argumentation, situated guesswork, embedded imagining, multiple perspectives, and innovative uses of everyday words to construct new meanings – that intersect in potentially productive ways with scientific practices. As students engage in scientific inquiry and argumentation, teachers can identify intersections between students' everyday knowledge and scientific practices, and use these intersections as the basis for instructional practices. When their cultural and linguistic experiences are used as intellectual resources, students with limited science experience or from nonmainstream backgrounds are able to conduct scientific inquiry and to appropriate scientific discourse as members of a science learning community.

While initially the research focused primarily on scientific inquiry and argumentation, it has evolved to a consideration of the role of students' first language in scientific sense making. The research considers students' first language in two senses (Ballenger, 1997; Warren et al., 2001). The first refers to students' mother tongue, such as Haitian Creole or Spanish. The use of students' first language allows for greater ease of communication and serves as a resource for their learning. Students' deep knowledge of their first language and their facility with its syntax and vocabulary allow them to refine distinctions and to express subtle nuances of meaning. The second meaning refers to what S. B. Heath (1983) termed "ways with words," that is, community-based discourse patterns that constitute the vehicle for children's initial language socialization. Even among individuals who speak the same language, linguistic and social practices, such as storytelling and argumentation, may take different forms and contexts. Thus, students come to school with varying levels of familiarity with the ways with words, for example, what kinds of discourse are appropriate and when, what kinds of arguments are allowed, and so on.

Sociopolitical Process of Science Instruction

Several studies have found that science instruction often reinforces power structures that privilege mainstream students, and that the substandard performance of other students may be due to their active resistance to science instruction and to schooling more generally. These studies generally rely upon ethnographic research and the collection of students' own narratives of their school experience.

M. T. Hayes and D. Deyhle (2001) used ethnographic methods to describe how science instruction was provided differently at two elementary schools, one serving predominantly middle-class White students and the other serving predominantly low-SES, ethnic minority students (ethnic breakdown is not presented in the article). At the first school, science instruction was fixed and rigid, and teachers emphasized conceptions of academic success that included raising standardized test scores, ensuring future academic performance, and going to college. At the second school, science instruction was more open, supportive, and personally relevant to students. However, while the latter group of teachers emphasized student engagement and enjoyment in science, they did not have specific or well-formed visions of how to prepare their students for standardized tests or for the rigors of future academic settings. The researchers surmise that although the latter type of science instruction might be perceived as better or more effective according to the current conceptions of science instruction, differential curricula and pedagogy between the two schools might be simply reconfigurations of social reproduction mechanisms based on existing racial/ethnic, socioeconomic, and political hierarchies.

A. Gilbert and R. Yerrick (2001) described the beliefs and practices of eight students and their teacher in a lower-track earth science class at a rural high school. The teacher (a White male) had expertise in physics, earth science, and environmental science. He selected 8 students from his class of 28 to represent the mix of students in any given section of the earth science course. He chose three Black females, three Black males, one White (Anglo) male, and one Cuban American male. These 8 students participated in weekly focus-group interviews, as well as individual interviews throughout the study. The results of this ethnographic study indicate that the quality of science instruction was subverted through a process of negotiation between teacher and students, in a context of low expectations and an unsupportive school culture. The teacher appeared to be concerned for his students and trying to do what he thought best for them, but he was bound by his lack of practical knowledge and experience in teaching science to nonmainstream students. Students were unable to separate the teacher from the larger dominant system he represented, and thus rejected him, disrupted lessons, provoked disciplinary action, and challenged his authority in ways that asserted their own separate identities. Student apathy and resistance, as well as teacher frustration and hostile language, reinforced the social distance between the students and the teacher and other school officials. The researchers argued that the existence of the lower-track science class is tied to societal agendas for maintaining and reproducing socioeconomic stratification.

Z. D. Sconiers and J. L. Rosiek (2000) and K. Tobin (2000) present personal narratives of teaching science in inner-city secondary schools with predominantly African American students. Sconiers was a middle school science teacher who took up the role of researcher, whereas Tobin was a researcher who took up the role of high school teacher. In two separate studies in two different educational settings, the researchers described (a) how the school system was structured in ways that failed to connect the enacted curriculum to the interests and extant knowledge of these students, (b) how students resisted learning science or relating to their teachers personally, and (c) how teachers with the best intentions failed to teach science in ways that were relevant to students. While recognizing the challenges inherent in such teaching situations, the researchers offer insights for educating prospective and practicing science teachers of students who are regularly marginalized in science classrooms. Sconiers reflects on why he, as an African American male science teacher, had only marginal success with two African American male students who resisted schooling in general and science learning in particular. He advocates that meaningful, authentic connections between school science and traditionally underserved students need to be made. Tobin suggests that science teachers consider such options as enacting multiple activities in each lesson, encouraging alternative ways of participating, setting up a portfolio system, and involving others (parents, siblings, guardians, and persons from the community) in supporting students' science learning.

Gilbert and Yerrick (2001), Sconiers and Rosiek (2000), and Tobin (2000) all describe mistrust of schooling, science instruction, and science teachers among those students who have traditionally been disenfranchised and oppressed by schooling in general and science education in particular. This mistrust can present a serious barrier to these students' science achievement, inasmuch as science inquiry requires scientific skepticism, a tolerance for uncertainty and ambiguity, and patience, all of which depend on a certain level of trust between teacher and students. Thus, science instruction, particularly inquiry-based instruction, is "trust intensive," and mistrust is exacerbated when science teachers do not expect students to succeed. The researchers argue that building trusting and caring relationships between teachers and students is necessary in order for students to take intellectual risks, which are in turn necessary in order to develop a real understanding of science content and practices.

English Language and Literacy in Science Instruction

So as not to fall behind their English-speaking peers, ELL students need to develop English language and literacy skills in the context of subject area instruction (August & Hakuta, 1997; García, 1999). Ideally, subject area instruction should provide a meaningful context for English language and literacy development, while advancing English skills provides the medium for engagement with academic content (Buxton, 1998; Lee & Fradd, 1998). In reality, however, ELL students frequently confront the demands of academic learning through a yet-unmastered language, without the kinds of instructional and institutional support they need. Furthermore, teachers often lack the knowledge and the institutional support necessary to address the complex educational needs of ELL students.

Science instruction typically has failed to help ELL students learn science in ways that are meaningful and relevant to them, while also failing to help them develop proficiency in oral and written English. An emerging body of literature indicates the positive impact of instructional interventions to promote ELL students' science knowledge and inquiry skills. Two areas of emphasis are found in the literature: (a) inquiry-based science instruction and (b) code-switching between students' home language and English.

Inquiry-Based Science Instruction

The studies described in this section, though relatively few in number, vary widely in terms of research questions, methods, and student outcomes. The studies ranged from small-scale descriptive research (Kelly & Breton, 2001) to large-scale intervention research (Amaral, Garrison, & Klentschy, 2002). Research methods ranged from an experimental design (Rodriguez & Bethel, 1983) to discourse analysis of classroom talk (Kelly & Breton, 2001). The studies examined various student outcomes: students' engagement in science discourse (Kelly & Breton, 2001), scientific writing (Merino & Hammond, 2001), and both science and writing outcomes (Amaral et al., 2002; Rodriguez & Bethel, 1983).

On the basis of observations of 57 randomly selected elementary classrooms serving predominantly Latino students, R. H. Barba (1993) found that most bilingual students, regardless of their program placement (transitional, bilingual, sheltered English, ESL, etc.), received instruction predominantly through teacher-directed expository instruction, rather than student-directed learning models or collaborative group work. Teachers in these classrooms often lacked proficiency in the children's native language (65% were monolingual English speakers), and the bulk of teacher talk was conducted in English. Culturally relevant examples, analogies, and elaborations were used far less than generic or mainstream elaborations in science instruction (3% vs. 97%).

Kelly and Breton (2001) examined how two bilingual elementary school teachers guided their students to engage in science inquiry through particular ways of framing problems, making observations, and engaging in spoken and written discourse. Drawing from the perspectives offered by classroom ethnography and anthropological studies of scientific practice, the researchers used a discourse analysis approach to examine the interactive processes that constructed science as disciplinary inquiry, at the same time that they constructed a community of students as scientists. While one teacher felt constrained about using Spanish because of external pressures, the other teacher regularly code-switched in her teaching. The results indicate that the processes of framing disciplinary knowledge and introducing students to conventionalized ways of observing, writing, speaking, and understanding required specialized discursive work on the part of the teachers. This work included engaging students in conversations through questioning, reframing ideas, varying use of languages, making reference to other classroom experiences, and devising interactional contexts for students to "talk science" under varying conditions.

Merino and Hammond (2001) examined how nine elementary school teachers facilitated bilingual students' learning of science concepts and skills through writing. The teachers implemented a science-based interdisciplinary approach in which a series of science inquiry lessons was integrated with other subject areas of the school curriculum. The instructional approach was grounded in what the authors refer to as "sheltered constructivism." First, students participated in activities under the guidance of teachers who contextualized tasks by using communicative techniques and students' home language; then, students followed up with further activities based on their own questions. Teachers were observed to use a variety of classroom activities and strategies to promote students' scientific writing. The researchers suggest that in addition to producing narrative texts (a common practice in elementary schools), elementary students should be provided with experiences in other genres of writing in content areas such as science. The researchers provide detailed information about genres of scientific writing, such as recording faithfully and in detail what transpires in the science lessons, maintaining careful records to understand what happens, reporting to others so that they can repeat an experiment, keeping a record of an experiment for future use, and providing visual representations of events. The researchers also provide detailed information about instructional strategies to promote scientific writing, such as visual models, student examples, models from the media or literature, video accounts, lab notes, and narrative journals.

Rodriguez and Bethel (1983) examined the effectiveness of an inquiry approach to science and language teaching aimed at developing classification and oral communication skills among bilingual Mexican American 3rd-grade students. From a population of 120 students in an urban elementary school in central Texas, a random sample of 64 students was selected for the experimental and control groups. These children had been participating in a bilingual education program since kindergarten, but as they learned English, they received an increasingly English-medium curriculum. They could understand, speak, read, and write both Spanish and English, but were not sufficiently fluent in English to participate in a regular English curriculum. (From the article, it is unclear whether science instruction was conducted in English or bilingually.) The researchers employed the Solomon Four-Group experimental design. For a period of 12 weeks, the experimental group participated in a sequential series of 30 science inquiry lessons that required manipulation of objects, exploration, and interaction with peers and the teacher. The children made observations and comparisons of familiar objects and then grouped them on the basis of perceived and inferred attributes. The control group participated in a science program that did not emphasize classification and oral communication skills specifically, nor did it require the manipulation of concrete objects. This program consisted of units developed by teachers working in the school district. Two test batteries to measure children's classification and communication skills were administered prior to and at the completion of the study. The ANOVA results indicated statistically significant improvements for the experimental group in both classification and oral communication skills, which the researchers ascribed to participation in science inquiry lessons.

The research by O. M. Amaral and colleagues (2002) examined the impact of a four-year intervention with elementary ELL students in a rural school district. The five areas of emphasis in this NSF-supported, district-wide local systemic reform initiative included high-quality curriculum, sustained professional development and support for teachers and school administrators, materials support, community and top-level administrative support, and program assessment and evaluation. Students in the district participated in kit- and inquiry-based science instruction that included the use of science notebooks. Although teachers and students had the freedom to use Spanish for facilitation of instruction, most instruction was in English in "bilingual" classes as well as sheltered/transitional English (now called structured English immersion) classes. All students were assessed in science using the Stanford Achievement Test that served as the statewide science assessment. Assessments in writing were conducted using the district writing-proficiency test. The assessment instruments were in English.

This study involved only those students who had been enrolled in the school district for the previous four years (during which the reform initiative gradually expanded to cover the entire district), including 615 students in 4th grade and 635 students in 6th grade. These students were divided into five groups based on the number of years (0–4) of their participation in the project. Once these groups were identified, they were further disaggregated by five linguistic categories, including limited English proficient, limited/fluent English speaking, fluent English proficient, English-only, and redesignated fluent English proficient. For some analyses, the five linguistic categories were combined into two (limited English proficient and English proficient). Student achievement in science and writing was compared (a) across the five levels of students' duration in the program (0–4 years) and (b) among five levels of English proficiency. A series of ANOVA tests were conducted for the two independent variables (duration in the program and English proficiency). The results indicate that with both 4th- and 6th-grade students, both science and writing achievement increased significantly in proportion to the number of years they participated in the program. Among both 4th- and 6th-grade students, English-proficient students performed significantly better than limited English-proficient students in both science and writing.

The research conducted by Amaral et al. (2002) has several unique features: (a) it is one of only a few studies that examined the impact of a district-wide reform initiative on student achievement; (b) student outcome measures included both science and writing; and (c) student outcome data were disaggregated by number of years of participation in the program (i.e., longitudinal impact) and level of English proficiency. However, it is difficult to infer mechanisms that led to increased student achievement gains, due to the lack of a control or comparison group, failure to assess students' science knowledge independently of their English literacy skills, and failure to take into account students' prior levels of literacy in the home language. The study suggests that district-wide implementation of a comprehensive reform initiative involving science curriculum, professional development, and classroom instruction presents challenges in terms of research design, management and analysis of data, and interpretation of the results with regard to causality of the intervention.

Code-Switching

One set of studies looked specifically at code-switching (the practice of alternating between two or more languages or language varieties within a single speech event) in science classrooms. Except for the studies by A. Luykx, O. Lee, and U. Edwards (in press) and M. E. Blake and M. V. Sickle (2001), which were conducted in the United States, all the others were done in African countries, using various qualitative research methods.

The code-switching examined by Luykx, Lee, and Edwards (in press) was less a conscious pedagogical strategy for promoting scientific discourse among ELL students than a logical result of constraints imposed by the policy context surrounding the school in question. The researchers conducted discourse analysis of science lessons in a combined 3rd- and

4th-grade class of beginning ELL students. The teacher was a monolingual English speaker, assisted by a bilingual co-teacher whose role consisted primarily of providing concurrent translation (in Spanish) of the teacher's English-medium instruction. The study contrasted classroom discourse from a regular class (with both teachers present) with discourse from a nontypical class in which the bilingual co-teacher was absent. The results indicate that the language ideologies underlying school policy and practice viewed different languages as essentially equivalent, neutral codes, and viewed science concepts as essentially independent of the language in which they are constructed or expressed. Analysis of the co-teacher's attempts to render the English-based science content into Spanish, and of students' attempts to negotiate the language barrier during class discussions of that content, demonstrated that science concepts are, in fact, tightly tied to the language in which they are constructed.

The researchers also examined how the use of concurrent translation in instruction shaped students' opportunities to engage in scientific discourse. In the regular class, students passively awaited the co-teacher's translations, and semantic discrepancies between the two languages went unresolved. Paradoxically, pedagogical conditions were more conducive to the development of scientific discourse when the bilingual co-teacher was absent, in that students took on a greater role in the construction of scientific understandings, actively negotiating meanings with the teacher and with one another. However, pressure to help the teacher understand their discussions, and the teacher's limited ability to understand and make herself understood by students, limited students' learning opportunities in this context also. The researchers conclude that concurrent translation is an ineffective pedagogical strategy, both with regard to teaching science and to helping ELL students develop English proficiency. Had the instructional situation lived up to its designation of "Curriculum Content in the Home Language," students would have had more opportunities to develop their scientific discourse. On the other hand, if the teachers had provided actual "sheltered English immersion" (or, alternatively, bilingual instruction), students would have had more opportunities to develop their English proficiency. The research suggests that it may not always be possible to effectively combine these two goals in a single lesson or activity.

Blake and Sickle (2001) worked with African American high school students on one of the Sea Islands in South Carolina. Their purposeful sample included students who demonstrated dialect diversity, retention in special education, consistent failure to pass the various sections of the South Carolina Exit Exam (SCEE), and little or no coursework in mathematics and science. The research aimed to help these students code-switch between their island dialect and Standard English so that they could communicate their understandings of science concepts and eventually pass mainstream courses and the SCEE. The study started with one student and expanded to nine students by the third year. Students were provided with a hands-on inquiry science curriculum, which was complemented by various techniques designed to promote their language development, such as focused discussions of science topics, conversations about the difference between fact and inference, a writing workshop, dialogue journals, and other "language experience activities" based on students' personal experiences in and out of the classroom. Case studies of two students indicated that consistent language interactions with the research team helped them become more sensitive to the language of the school and of the test. When they improved their ability to code-switch from the highly inferential local dialect to a more explicit and detailed Standard English, they improved their science achievement. Both students gained over 100 points in the reading, writing, and mathematics sections of the SCEE and were able to graduate from high school. The researchers stressed the importance of teacher preparation programs that expose future teachers to different cultures and dialects, as well as the value of promoting students' code-switching abilities, rather than attempting to eradicate their nonstandard dialects.

Similarly positive effects of code-switching in science education contexts were observed in African countries. As part of a larger project investigating the role of language in conceptual change, M. Rollnick and M. Rutherford (1996) examined code-switching between SiSwati and English by primary teacher trainees in Swaziland as they carried out an experiment on air pressure. Qualitative analysis of the transcripts of audiotapes during the trainees' group work (50 episodes representing seven groups from three different colleges) indicated that code-switching norms varied widely from one group to another. This variability was closely related to the social structures that developed in the groups as they carried out the experiment, and also to the degree to which the groups were normally restricted to English in their regular classes. The findings suggest that the use of SiSwati served several important functions, such as articulating existing ideas, clarifying concepts, eliminating misconceptions, and formulating new ideas. The researchers conclude that extensive probing in students' home language is required if the teacher is to identify students' alternative conceptions. They also note that bilingualism can be an advantage in concept acquisition, as it gives learners experience with different representations of the same idea.

Through an ethnographic study of science instruction in three rural elementary schools in Kenya, A. Cleghorn (1992) found that science content was made more accessible to students when teachers incorporated use of local languages (Kikuyu, Luo, and to a lesser degree Kiswahili) in a variety of code-switching patterns, rather than adhering strictly to the schools' policy of English-only instruction. Meaningful instruction in English was hampered not only by students' limited familiarity with English but also by teachers' own inability to use English spontaneously, although they were fluent enough to use it for verbatim transmission of content and rote drill of closed questions. In general, instruction that utilized code-switching was clearer than instruction that relied exclusively on either English or the local language. Use of local languages along with English provided a means for drawing on students' first-language skills in the construction of meaning, linking the foreign cultural content of instruction to students' experiences outside of school, and connecting the concrete to the abstract. The results suggest that given the limited knowledge of English among Kenya's general population, it may not currently be possible to combine English language development with the effective teaching of subject content. Cleghorn stresses the importance of sociolinguistic study of the instructional process itself for educational planning, teacher training, and other educational reforms linked to language policy and national development. Cleghorn also notes the considerable evidence indicating that purposeful maintenance of students' first language assists in the development of literacy skills in the target second language.

Setati and colleagues (2002) described how primary and secondary teachers and students in 10 urban and rural schools in South Africa moved from informal, exploratory talk in students' respective home languages to discourse-specific talk and writing in English. The study involved 25, 23, and 18 teachers (including mathematics, science, and English teachers) during 1996, 1997, and 1998, respectively. Data from multiple sources (classroom observations and videotapes, teacher interviews, teachers' narratives of specific lessons, questionnaires, and examples of student work) were analyzed for evidence of teachers' and students' code-switching practices. The results indicate that few teachers and students were able to successfully "complete the complex journey" from informal, exploratory talk in the vernacular to discourse-specific talk and writing in English. Code-switching practices differed by region, grade level, and subject being taught; the researchers note that these important differences are often concealed by analyses focusing on overall patterns across teachers. While South African language policy officially advocates additive bi-/multilingualism, the standard practice of assessing students exclusively in English and the fact that rural students' only exposure to English is in school put pressure on teachers to use English as much as possible. Code-switching is thus perceived as a "dilemma" even though teachers feel the need to do it. The researchers also caution that the utility of having students learn subject matter content through the medium of vernacular languages is substantially limited if they do not also learn to talk in the formal, English-based discourses of mathematics and science.

Discussion

Research on science instruction with nonmainstream students has been conducted from a wide range of perspectives, drawing upon anthropology, sociolinguistics, cognitive science, and critical theory. Despite this theoretical diversity, most studies employed classroom observations of instructional processes as a primary research method. The Rodriguez and Bethel (1983) study was the exception, employing an experimental design based on test batteries. Also, most studies used students' participation and engagement in science classrooms as outcome measures, but did not report achievement data. The exception is the study by Amaral and colleagues (2002), which measured achievement in both science and literacy (writing).

Different researchers have proposed different approaches to science instruction based on their particular theoretical/conceptual perspectives. Research on culturally congruent instruction suggests that when students do not share the "culture of power" of the dominant society (e.g., Western science), teachers need to make that culture's rules and norms explicit and visible so that students can learn to cross cultural borders between their home environment and the school environment. For students with limited science experience or those who come from backgrounds in which questioning and inquiry are not encouraged, teachers can move progressively along the teacher-explicit to student-exploratory continuum, to help students learn to take the initiative and assume responsibility for their own learning.

In contrast, research on cognitively based science instruction suggests that teachers need to understand the complex dynamics between scientific practices and students' everyday knowledge. As teachers identify and incorporate students' cultural and linguistic experiences as intellectual resources for science learning, they provide opportunities for students to learn to use language, think, and act as members of a science learning community.

Research from the sociopolitical perspective suggests that science instruction is influenced by power structures that privilege mainstream students. Teachers need to build trusting relationships with students who have been marginalized and disenfranchised in science classrooms and to provide safe environments for students to take part in learning science. Unless students see science as personally meaningful to their current and future lives, they are likely to disengage or actively resist learning science.

The studies on science instruction with ELL students in the United States consistently focus on hands-on, inquiry-based instruction. Such instruction provides opportunities for ELL students to develop scientific understanding, engage in inquiry, and construct shared meanings more actively than does traditional textbook-based instruction, for a number of reasons: (a) hands-on activities are less dependent on formal mastery of the language of instruction, thus reducing the linguistic burden on ELL students; (b) collaborative, small-group work provides structured opportunities for developing English proficiency in the context of authentic communication about science knowledge; and (c) hands-on activities exploring natural phenomena make science concepts more accessible to students with limited science experience than do approaches based on decontextualized textbook knowledge. By engaging in science inquiry, ELL students develop their English grammar and vocabulary, as well as their familiarity with scientific genres of writing. Furthermore, inquiry-based science instruction promotes students' communication of their understanding in a variety of formats, including written, oral, gestural, and graphic (Lee & Fradd, 1998; Rosebery et al., 1992).

With the exception of the study by Setati and colleagues (2002), those on code-switching indicate that alternating between English and the home language allows ELL students (and teacher trainees, in the case of Rollnick and Rutherford, 1996) to engage science content and process more deeply. The literature from both Africa and the United States suggests that policies mandating the exclusive use of English in science instruction are out of step both with research-based notions of "best practice" and with students' own strategies for negotiating content in a new language. Given the growing number of U.S. classrooms in which ELL students share the same home language (usually Spanish), code-switching as part of students' science learning is probably much more common than is generally realized, school language policies notwithstanding. Rather than attempt to force students' linguistic practice into the mold mandated by English-only policies, researchers and educators might direct more attention to code-switching and the ways in which it influences science learning with ELL students.

6

Science Assessment

There is an extensive body of literature on educational assessment in general, and also a large body of literature on assessment with nonmainstream students, including ELL students. However, research on science assessment with nonmainstream students (both large-scale and classroom assessment) is extremely limited (Lee, 1999a; Solano-Flores & Trumbull, 2003). As discussed earlier (see "Accountability as the Policy Context for Science Education" in Chapter 2), science is often not part of large-scale or statewide assessments, and even when it is tested, science usually does not count toward accountability measures. Additionally, because assessment of ELL students tends to concentrate on basic skills in literacy and numeracy, other subjects such as science tend to be ignored. Given that science is not part of accountability, research on science accommodations with ELL students is sparse. The exact number of states that assess science and/or include science in accountability measures is constantly changing as more states implement statewide science assessments. This trend is likely to intensify, as science will be part of the No Child Left Behind Act starting in 2007.

A critical issue concerning valid and equitable assessment in multicultural and multilingual settings is how to address cultural and linguistic influences on students' measured performance. A small number of studies about science assessment are divided into two categories: (a) science assessment with culturally diverse student groups and (b) science assessment with ELL students. Since these studies address a range of issues about science assessment, there are often only a few studies on a particular issue. Given the limited empirical research on science assessment with nonmainstream students, it is unclear whether new assessment technologies and innovations present more hopes or obstacles to these students.

Science Assessment with Culturally Diverse Student Groups

One way to promote equitable assessment is to make assessments relevant to the knowledge and experiences of diverse student groups in their home and community environments. This approach focuses on the content of science assessments, which has traditionally made few connections to non-mainstream students' lives. This is partly due to the fact that relatively few teachers or test developers have an in-depth knowledge of nonmainstream students' cultural beliefs and practices. Additional difficulties involve the presence of students from different cultural backgrounds and degrees of assimilation to the mainstream within the same classroom.

G. Solano-Flores and S. Nelson-Barber (2001) propose the notion of "cultural validity" to address sociocultural influences that shape student thinking and the ways in which students make sense of and respond to science items. These sociocultural influences include the values, beliefs, experiences, communication patterns, teaching and learning styles, and epistemologies originating in their home communities, as well as the socioeconomic conditions in which they live. Furthermore, students of differing cultural backgrounds may have alternative ways of expressing their ideas, which may mask their knowledge and abilities in the eyes of teachers unfamiliar with students' linguistic and cultural norms. Grounded on cross-cultural studies on science learning and assessment, the researchers identify five areas in which the notion of cultural validity can contribute to the improvement of science assessment: (a) student epistemology, (b) student language proficiency, (c) cultural worldviews, (d) cultural communication and socialization styles, and (e) student life context and values. The researchers suggest that "ideally, if cultural validity issues were addressed properly at the inception of an assessment and throughout its entire process of development, there would be no cultural bias and providing accommodations for cultural minorities would not be necessary" (p. 557).

The areas identified by Solano-Flores and Nelson-Barber are addressed (though sliced rather differently) in a study by this book's authors and their colleagues (Luykx, Lee, Mahotiere, et al., in press), which examined cultural and home language influences on students' written responses on paper-and-pencil science assessments. The assessment instruments were part of a larger project (see Lee et al., 2005) involving an inquiry-based science curriculum for 3rd- and 4th-grade students (White, Hispanic, African American, and Haitian American) in a large urban school district. The instruments measured students' science knowledge and inquiry skills on two curriculum units per grade: Measurement and Matter for 3rd grade, and the Water Cycle and Weather for 4th grade. The results revealed numerous types of linguistic and cultural influences in students' responses. Though the number of examples was small, relative to the entire data set, they displayed features that could conceivably skew assessment of individual students' science knowledge and inquiry skills. These features included non-standard spellings of English words, which reflected the phonology of students' home language (and were often unintelligible to adult readers unfamiliar with Spanish); semantic confusion around science terms with more than one possible meaning (e.g., *state, gas,* and *record*); genre confusion regarding the interpretation of scenario-based science questions; and responses reflecting the practices and interpretive frameworks of students' home environments, rather than those assumed by the test developers/research team. The results also indicated that the assessment instruments were shaped by the test developers' own cultural assumptions and linguistic practices, to a greater degree than the research team had realized.

On the basis of the results, Luykx, Lee, Mahotiere, and colleagues (in press) draw several conclusions. First, assessment instruments are inevitably cultural products, and while attempts to avoid obvious cultural bias are laudable, they can never achieve complete cultural neutrality. Second, many of the cultural and linguistic influences present in students' responses are unintelligible to scorers unfamiliar with students' home language and culture. For teachers of ELL students, students' lack of English proficiency or limited familiarity with mainstream culture may masquerade as lack of science knowledge or inquiry skills. Finally, attempts to ground assessment items in students "real-world experience" may be misguided, given that experience differs according to students' cultural and linguistic backgrounds. More equitable assessment practices might involve avoiding assessment items' reliance on cultural knowledge and experiences that are assumed (sometimes mistakenly) to be shared by all students, in favor of designing assessments that require only knowledge of science content taught in school. However, this approach would require that all students have equal access to quality science instruction in school, which is often not the case, especially for ELL students.

Another way to promote valid and equitable assessments is to determine more effective formats for assessing student achievement. Traditional multiple-choice tests have been criticized for failing to measure the types of knowledge, abilities, and skills that science students should be expected to learn (AAAS, 1989, 1993; NRC, 1996). Instead, alternative (or performance) assessments are called for, including open-ended or essay items, laboratory-based practical tests, portfolios, and opportunities to design and conduct experiments or projects (Ruiz-Primo & Shavelson, 1996). One of the most important issues in using alternative assessments is their fairness to different student groups. Fairness in this context means "the likelihood of any assessment allowing students to show what they understand about the construct being tested" (Lawrenz, Huffman, & Welch, 2001, p. 280). Given the limited research on alternative science assessments with nonmainstream students, both advocates and critics have based their claims on inferences and insights drawn from related research endeavors, rather than on empirical studies that directly address school science (see the discussion in Lee, 1999a). Furthermore, existing studies on science assessments show inconsistent results, for example, with regard to the effect of assessment formats on science outcomes of students from different racial/ethnic groups (Klein et al., 1997; Lawrenz et al., 2001).

S. P. Klein and colleagues (1997) examined whether the differences in mean scores among racial/ethnic (and gender) groups on science performance assessments are comparable to the differences that are typically found among these groups on traditional multiple-choice tests. The research, which involved more than 2,400 students in grades 5, 6, and 9 from 90 classrooms across 30 schools, was part of a field test of California's statewide testing program, the California Learning Assessment System (CLAS). Students completed several hands-on science performance assessments and took multiple-choice items. Additionally, 5th- and 6th-grade students took the multiple-choice science subtest of the Iowa Tests of Basic Skills (ITBS). The testing sessions were conducted with students individually. The results indicate that differences in mean scores among racial/ethnic groups were not related to test formats (i.e., performance assessment vs. multiple-choice items) or question types within performance tasks (e.g., "interpretation" questions vs. "analysis" questions). Regardless of test format or question type, Whites and Asians had significantly higher mean scores than Blacks or Hispanics. These results remained consistent when other factors were taken into account, such as variations among schools or teachers. The authors conclude that changing test format or question type is unlikely to have much effect on the differences in mean scores among racial/ethnic groups.[1] These results should be interpreted with caution, since the researchers did not control for other student characteristics, such as SES or parents' educational level, which are likely to influence students' academic performance and reflect racial/ethnic divisions to some degree.

F. Lawrenz and colleagues (2001) examined science achievement outcomes for different subgroups of students using different assessment formats. A nationally representative sample of approximately 3,500 9th-grade science students from 13 high schools throughout the United States completed a series of science assessments designed to measure their level of achievement on the national science education standards. All of the schools were using a curriculum designed to meet the *National Science Education Standards* (NRC, 1996). Teachers from each school were involved in the curriculum development, and the resulting materials were provided to all participating classrooms. All assessment items were selected from

---

[1] On the other hand, changing the test format or question type had statistically significant effects on the differences in mean scores between boys and girls.

existing sources, such as the NAEP, International Assessment of Educational Progress (IAEP), and Second International Science Study (SISS). The assessments included a multiple-choice test, an open-ended written test, a hands-on lab skills test, and a full hands-on investigation. The results showed that different assessment formats measured different competencies, particularly the hands-on formats in comparison to the multiple-choice and open-ended formats. In contrast to the results of the study by Klein and colleagues (1997), these results showed that the achievement of students from different racial/ethnic groups varied by assessment format. The typical trend of lower achievement for African American and Hispanic/Latino students and higher achievement for Asian American and Anglo students was found in all assessment formats. Interestingly, there were switches in relative order within the two top and the two lower performing groups on the hands-on tests. These results suggest that using different assessment formats may affect the science outcomes of students from different racial/ethnic groups. These results should be interpreted with caution, since like Klein and colleagues (1997), the researchers of this study did not control for other student characteristics, such as SES or parents' educational level.

Science Assessment with ELL Students

Assessment of ELL students is complicated by such issues as which students should be included in accountability systems, what assessment accommodations will allow them to demonstrate their knowledge and abilities, and how content knowledge can be assessed separately from English proficiency or general literacy (Abedi, 2004; Abedi et al., 2004; August & Hakuta, 1997). Research on these issues in science assessment is very limited.

Most research on linguistic factors in assessment of ELL students has focused on the effectiveness of various testing accommodations (e.g., use of bilingual dictionaries or subject-specific glossaries, extra time to complete assessments) (Abedi, 2004; Abedi et al., 2004; O'Sullivan et al., 2003). Although assessments can be made more comprehensible to ELL students by avoiding unnecessarily complex grammatical constructions, polysemic terms (i.e., terms with more than one meaning), and idiomatic expressions, such accommodations are not regularly employed and may not reflect the type of English used in instruction.

In terms of assessment accommodations, the 2000 NAEP report is the first since its inception in 1969 to include results for students with disabilities (SD) and limited English proficiency (LEP) (O'Sullivan et al., 2003). Two sets of results are reported: "accommodations-permitted" and "accommodations-not-permitted." Accommodations included, but were not limited to, one-on-one testing, small-group testing, access to bilingual dictionaries, extended time, reading aloud of directions, recording of students' answers by someone else, signing of directions (for deaf students), and use of magnifying equipment and large-print books (for visually impaired students). At grade 4, the accommodations-permitted results, which included slightly more SD and LEP students because of the availability of accommodations, were two points lower than the accommodations-not-permitted results, and this difference was statistically significant. At grades 8 and 12, there was no statistically significant difference between the two sets of results. Unfortunately, the results are not disaggregated by SD or LEP separately, because of the small numbers of SD and LEP students who were assessed at each grade level, with or without accommodations. The results are also confounded by the fact that the accommodations-permitted group included slightly more SD and LEP students.

Assessment for ELL students should ideally distinguish science knowledge from English language proficiency, although this is rarely done in research and assessment programs. J. M. Shaw (1997) examined the use of science performance assessment with two classes of ELL students in a high school in a large metropolitan area in northern California. The research setting presented an ideal site for a best-case implementation of a performance assessment with ELL students. The school implemented bilingual education programs with extensive human and material resources for effective instruction of ELL students. The two teachers participating in the research had experience and training in the teaching of hands-on inquiry science to ELL students. One was bilingual with Spanish as her native language, and the other a native speaker of English who was conversant in Spanish. This research was conducted using multiple data sources over a 16-month period in the two science classes.

The study (Shaw, 1997) focused specifically on a four-day performance assessment task in all five consecutive periods of sheltered science instruction in the two classes taught by the two teachers. The assessment task on heat energy involved an open-ended inquiry and hands-on investigation by students working in small groups. Student responses on assessment items were scored by the researcher and the two teachers using a scoring rubric. Both qualitative and quantitative results were presented with respect to three reference points: the student perspective, the teacher perspective, and analysis of test scores. Both students and teachers were generally in favor of using performance assessment. Students felt that it was a valuable learning experience and an accurate measure of their scientific knowledge and skills. Teachers saw a direct connection between performance assessment and their instructional practice. Both students and teachers, however, expressed concerns and gave suggestions for improving the assessment. ANOVA was conducted with students' test scores to determine the degree to which the assessment functioned as a measure of science knowledge or English language proficiency. Only the inquiry procedure, the most text-dependent item scored, was significantly affected by students' level of English proficiency. Conversely, graphs, calculations using an equation and a data table, and final summary questions were significantly affected by students' level of science knowledge. Thus, there was no simple answer to the question of whether performance assessments accurately measure ELL students' science knowledge; instead, the answer depends on the assessment task in question. Furthermore, student performance was affected by other variables, such as small-group interactions among students and differing levels of specificity and elaboration provided by the teachers for their students.

To address the complexities of linguistic and cultural factors in assessment more effectively, Solano-Flores and E. Trumbull (2003) propose a new paradigm to ensure valid and equitable assessment of ELL students and other nonmainstream students. Most current efforts focus on assuring test validity by attempting to eliminate the confounding effects of nonmainstream students' language and culture. Under the proposed new paradigm in assessment of ELL students, assessment efforts should be oriented in the opposite direction. It is virtually impossible to construct tests that are free from cultural and linguistic influences. Therefore, understandings of mainstream and nonmainstream languages and cultures must guide the entire assessment process, including test development, test review, test use, and test interpretation. One approach would be to design assessments in students' home language. Although this raises issues of validity relative to the English-language versions, it should be weighed against the threats to validity inherent in not testing students in their home language. On the other hand, assessing students in the home language may not give an accurate picture of their content knowledge if instruction has been carried out in English or if students' literacy skills in the home language have not been developed. Giving ELL students the same items in *both* English and their native language has the potential to produce more fine-grained understandings of the interactions among first- and second-language proficiency, students' content knowledge, and the linguistic and content demands of test items. Research in this vein has illustrated the proposed paradigm using science topics assessed in English, Chinese, Haitian Creole, and Spanish (Solano-Flores et al., 2001). However, this perspective has so far gained little ground in policy and assessment circles within the U.S. educational system.

As high-stakes assessment and accountability across subject areas become more prevalent, questions arise with regard to the consistency of large-scale and classroom assessments. S. H. Fradd and O. Lee (2000) addressed the challenges involved in combining standardized and informal assessments for ELL students learning science. They worked with all 13 4th-grade teachers at two elementary schools with high proportions of Spanish-speaking students. The students in these classrooms participated in an instructional intervention that involved the two project-developed instructional units and teacher professional development. Dependent t-tests were used to compare the pretest and posttest results after each year of the project. Science achievement scores on the two unit tests showed that students performed significantly better at the end of the school year compared to the beginning of the year. However, assessments of students' writing (for both science content and English language conventions) showed discrepancies in their performance between standardized and informal assessments. Students who performed successfully on statewide writing assessments employing a scripted five-paragraph composition format had difficulties going beyond this formula when asked to describe and make predictions (in writing) about a three-week-long science project. The researchers emphasize the need to integrate informal assessments that can provide insights into students' learning strengths and needs with standardized assessments that can establish benchmarks toward which all students must strive.

Discussion

Given the limited research, it is difficult to draw conclusions about how to ensure valid and equitable science assessment with nonmainstream students. Both advocates and critics of alternative assessment or accommodation strategies have based their claims mostly on inferences and insights drawn from related research endeavors, rather than from empirical studies on school science assessments per se (Lee, 1999a; Ruiz-Primo & Shavelson, 1996). Furthermore, those few existing studies that do address science assessment show inconsistent results (e.g., Klein et al., 1997; Lawrenz et al., 2001).

Although there has been considerable debate and discussion on the question of cultural bias in academic assessments, not all educators are convinced that cultural bias is a problem. Some are apt to attribute the differential academic performance of diverse students groups to deficiencies in students' home environments or cognitive abilities, rather than to cultural bias in assessment practices. Among those who do give credence to claims of cultural bias, there seem to be two opposing perspectives on assessment of students from nonmainstream backgrounds. The first aims to ensure test validity by removing cultural bias from assessment instruments and practices. In contrast, a recently emerging approach advocates that nonmainstream students' cultural beliefs and practices be incorporated throughout the assessment process (Solano-Flores & Nelson-Barber, 2001).

Although the latter approach may potentially solve some problems regarding cultural bias, it presents its own challenges. First, test developers and teachers often do not know enough about nonmainstream students' communities of origin to design assessments that are relevant to students' lives outside of school. Second, attempts to represent another culture by those who are not participating members of that culture always run the risk of stereotyping. Third, many schools contain students from numerous cultural backgrounds, so that making assessment instruments "culturally relevant" to some students may make them culturally inappropriate for others. Fourth, the notion of using different versions of assessment instruments with different student groups raises obvious problems with regard to validity across groups. Finally, even assuming that these other issues might be resolved, the unmistakable trend toward large-scale standardization of academic assessments means that the political and institutional support for widespread tailoring of assessments to specific cultural groups is unlikely to be forthcoming.

Many students from nonmainstream cultural backgrounds are also in the process of acquiring English as a new language. For these students, cultural bias is likely to be compounded by communicative difficulties. For ELL students who are assessed in a language they have not yet mastered, there is no easy solution to the problem of validity and equity. The science concepts, discourse, and assessment practices commonly used in U.S. schools are inextricably tied to the usage of American English. Thus, until students have mastered that language (which generally takes quite a bit longer than the one or two years of ESOL instruction they receive), assessing them in English cannot be assumed to provide an accurate picture of their science knowledge. On the other hand, assessing ELL students in their home language raises problems of validity, resources, and compatibility with the language of instruction (Solano-Flores & Trumbull, 2003). Furthermore, even when items are administered to ELL students in both English and the home language, ensuring the comparability of assessment instruments in the two languages is complicated.

In the current policy context of standardization, high-stakes assessment, and accountability, designing and implementing assessments for specific cultural and linguistic groups would be not only expensive and politically unpopular but also open to psychometric and other technical problems (Abedi, 2004; Abedi et al., 2004; Solano-Flores & Trumbull, 2003). For ELL students, possible solutions to this dilemma are further constrained by the spread of "English-only" legislation that prioritizes students' acquisition of English over their subject area knowledge (Abedi, 2004; Abedi et al., 2004; Gutiérrez et al., 2002). In light of all of these challenges, valid and equitable assessment of nonmainstream students remains one of the thorniest difficulties in educational policy and practice.

CREATING EQUITABLE LEARNING ENVIRONMENTS

To create equitable learning environments for nonmainstream students, teachers' knowledge, beliefs, and practices should evolve throughout their professional careers, continually incorporating our emerging understandings of student diversity and its effects on learning. Additionally, policies and practices at the state, district, and school level should be informed by these understandings in order to provide the necessary support to create such environments. When policies and practices at any level of the education system fail to provide support, teachers (and researchers) face difficulties in promoting students' science learning. Furthermore, there is a growing realization among educators that school science should be closely connected to the knowledge and experiences that nonmainstream students have acquired in their homes and communities. Although the majority of studies were conducted in classroom settings, a few examined specific features of students' home and community environments and their connections to school science. In this section, we present research results in the following areas: (a) science teacher education, (b) school organization and educational policy, and (c) home/community connections to school science.

7

Teacher Education

In contrast to the growing diversity among the student population, the teaching profession is increasingly dominated by White female teachers. O. Jorgenson (2000) states that "school districts across the United States confront an urgent shortage of minority educators, while the number of minority students in the public schools steadily increases. This imbalance is expected to worsen" (pp. 1–2). M. Haberman (1988) further states that "[h]aving too few minority teachers is merely one manifestation of under-educating minority children and youth in inadequate elementary and secondary schools" (p. 39).

Teachers need not come from the same racial/ethnic or linguistic background as their students in order to teach effectively (Ladson-Billings, 1994, 1995). Given the increasing student diversity even within individual classrooms, matching teachers with students of similar backgrounds is often not feasible. But when teachers of any background are unaware of the cultural and linguistic knowledge that their nonmainstream students bring to the classroom (Gay, 2002; Osborne, 1996; Villegas & Lucas, 2002), or when they lack opportunities to reflect upon how students' minority or immigrant status may affect their educational experience (Cochran-Smith, 1995a, 1995b; Valli, 1995), there is clearly a need for teacher education that specifically addresses teachers' beliefs and practices with regard to student diversity as it relates to subject areas.

Teachers face the challenge of making academic content accessible and meaningful for students from a broad range of cultural and linguistic backgrounds. To meet this challenge successfully, teachers must be equipped with knowledge of academic content and processes, ways in which academic content and processes may articulate with students' own cultural and linguistic knowledge, pedagogical strategies appropriate to multicultural settings, and awareness of how traditional educational practices have functioned to marginalize certain groups of students and limit their learning opportunities. In this chapter, three areas of research on teacher education are addressed: (a) teacher preparation, (b) teacher professional development, and (c) teacher education with regard to ELL students.

Teacher Preparation

In their review of the literature on prospective teachers' beliefs about multicultural issues, L. A. Bryan and M. M. Atwater (2002) discuss three categories of beliefs that they argue should be the focus of science teacher preparation programs aiming to meet the challenges of instructing an increasingly diverse student population: (a) student characteristics, (b) external influences on learning, and (c) appropriate teacher responses to diversity. Many prospective teachers believe that students from nonmainstream backgrounds are less capable than mainstream students. Teachers also tend to ascribe problems associated with students' learning to students' lives outside of school, rather than to teachers' beliefs and actions toward students in the classroom. Additionally, teachers are largely unaware of cultural and linguistic influences on student learning, do not consider "teaching for diversity" as their responsibility, purposefully overlook racial/ethnic and cultural differences, accept inequities as a given condition, or actively resist multicultural views of learning.

The literature review by Bryan and Atwater (2002) is largely based on the areas of elementary, reading, or language arts education, due to limited literature specifically addressing science or mathematics education. The researchers conclude that most prospective science teachers enter their teacher preparation programs with little or no intercultural experience and with beliefs and assumptions that undermine the goal of providing an equitable education for all students. Furthermore, many graduate without fundamentally changing their beliefs and assumptions, despite their experiences in teacher preparation programs. Thus, teacher preparation programs should provide prospective teachers with intercultural experiences that challenge their beliefs and assumptions about student diversity, if they are to learn to teach all students effectively.

An emerging literature on science teacher preparation supports the conclusions of Bryan and Atwater (2002). The studies discussed below employ case study or other qualitative approaches to examine science methods courses or teacher preparation programs. Many of them include prospective science teachers from mainstream backgrounds who encounter issues of student diversity for the first time in a university course or in a classroom setting. Overall, the results of these studies indicate that prospective science teachers have a hard time making fundamental or transformative changes in their beliefs and practices with regard to student diversity throughout their teacher preparation programs. Even when changes in teacher beliefs and practices occur, such changes are demanding and slow.

Science Methods Courses

One set of studies deals with science methods courses designed to help prospective science teachers foster positive beliefs and effective practices with regard to nonmainstream students. S. A. Southerland and J. Gess-Newsome (1999) worked with 22 prospective elementary teachers, almost all of them from mainstream backgrounds, in an elementary science methods course that used a variety of teaching methods to focus on issues related to inclusive science teaching. The results indicate that these teachers held a positivist view of knowledge, teaching, and learning. They believed that the goal of inclusive science teaching was to make a fixed and defined body of scientific knowledge accessible to all students, presumably within the confines of students' fixed abilities. They also believed that the goal was to help diverse learners think like mainstream students and to eliminate as much of the diversity as possible.

R. K. Yerrick and T. J. Hoving (2003) worked with prospective secondary science teachers, almost all from mainstream backgrounds, in a field-based secondary science methods course. The field experience involved working with predominantly rural Black high school students in a lower-track earth science class. Initially, all the prospective teachers designed, taught, and reflected upon their lessons from a relatively egocentric perspective, referring to their own experiences as learners as a guide to good teaching. All the teachers demonstrated similar practices, made similar inferences about teaching and learning, and relied on similar domains of knowledge to gauge their teaching. By the end of the course, two discrete categories of prospective teachers emerged: (a) those who demonstrated an ability to reflect on and revise their practices and to engage in the production of new pedagogical knowledge, and (b) those who deflected efforts to shift their thinking and instead reproduced their own educational experience with a new student population.

A. Rodriguez (1998b) worked with prospective secondary science teachers from mainstream backgrounds in a year-long science methods course. The course was based on a conception of multicultural education as integrating a political theory of social justice with a pedagogical theory of social constructivism. This approach aims to enable prospective teachers to teach for both student diversity (via culturally inclusive and socially relevant pedagogy) and scientific understanding (via critically engaging and intellectually meaningful pedagogy). The results showed promise in terms of assisting prospective teachers to critically examine their prior beliefs about what it means to be a successful science teacher. Most became aware of the importance of creating science classrooms where all students are provided with opportunities for successful learning. On the other hand, several teachers demonstrated strong resistance to ideological change, due to feelings of disbelief, defensiveness, guilt, and shame that Anglo-European prospective teachers often experience when they are asked to confront racism and their own racial privilege. They also demonstrated resistance to pedagogical change, due to the roles that they felt they needed to play in order to manage conflicting messages about what was expected of them, both from their cooperating teachers (i.e., cover the curriculum and maintain class control) and from their university supervisors (i.e., implement student-centered, constructivist class activities).

B. R. Brand and G. E. Glasson (2004) explored the development of belief systems in relation to racial/ethnic identities in the early life experiences of prospective teachers, and how their beliefs influenced their views about diversity in science classrooms and science teaching pedagogy. The study involved three prospective science teachers enrolled in a graduate licensure program, including an Asian male from a suburban setting, an African American male from an urban setting, and a White male from rural Appalachia. These three were selected because their racial/ethnic and cultural backgrounds made them less likely to become science teachers. As part of two secondary science teaching methods courses, the participants completed two internship placements in rural, suburban, and urban areas. The results of this ethnographic study indicate that the prospective science teachers were reluctant to embrace diversity because of racial/ethnic encapsulation and negative personal experiences in their own lives. Additionally, crossing cultural borders was threatening to them because accepting a new set of norms or beliefs could imply that something was wrong with their original beliefs. The results suggest that science teacher preparation programs should challenge and expand teachers' beliefs about racial/ethnic identities in order to promote self-efficacy and an awareness of the impact such beliefs have on science teaching.

In contrast to these studies describing challenges and difficulties that prospective science teachers experienced in their science methods courses, the study by E. V. Howes (2002) focused on the strengths that could assist teachers in developing an effective, inclusive science pedagogy. She worked with four prospective elementary teachers (two White and two African American) in an elementary science methods course. The strengths that emerged from these prospective teachers included a propensity for inquiry, concern for children, and an awareness of school/society relationships. In particular, the two African American teachers expressed a belief that schooling tends to work against social justice, a desire to use schooling to work for social justice, and a willingness to bring historical and cultural examples into the science classroom.

Student Teaching

Several other studies deal with student teaching in science teacher preparation programs. Consistent with the results of the studies conducted in science methods courses, these indicate that even those prospective science teachers committed to educational equity still face challenges related to student diversity during their student teaching.

J. A. Luft, J. Bragg, and C. Peters (1999) present a case study of a prospective secondary science teacher who was concerned with equitable instruction in her classroom. The study specifically examined her student teaching experience and the constraints she experienced as an Anglo teacher with predominantly Hispanic students and a small number of Native American and African American students. The teacher experienced: (a) unfamiliarity with her students and their life experiences, (b) marginalization by her students and colleagues as she tried to create new science lessons for students, and (c) a desire to make her science instruction more relevant to her students. Some of the difficulties within each of these three areas were resolved, while others remained present throughout her student teaching experience. The results revealed the complexity of learning to teach in a school where most students came from cultural backgrounds different from the teacher's own background.

J. A. Luft (1999) interpreted these same results in terms of cultural border crossing, with regard to cultural borders between: (a) the teacher and her Hispanic students, (b) the teacher's instructional philosophy and that of the other teachers at the school, and (c) the teacher and the school culture. While some of the teacher's efforts at border crossing were successful, others were not. This study suggests that prospective teachers encounter multiple cultural contexts, some consistent and some inconsistent with their instructional philosophy. Teacher educators need to recognize the cultural borders that prospective teachers will encounter when working with diverse student groups, and encourage them to examine their beliefs about teaching and learning as a means to acknowledging and understanding these borders.

L. D. Bullock (1997) designed a program to provide prospective science teachers with opportunities to examine their beliefs and practices regarding gender, ethnicity, and science education during student teaching. The researcher worked during one semester with six prospective secondary science teachers (their cultural and linguistic backgrounds are not specified in the article). Program activities revolved around four areas of focus: (a) equitable representation in curriculum materials, (b) equitable treatment within the classroom, (c) equitable opportunities in the laboratory setting, and (d) equitable evaluation of student performance. Although the student teachers approved of the program initially, they grew increasingly dismissive of issues of gender and ethnic equity once they began struggling with the inadequate academic preparation of their students and scant material resources. Eventually, the student teachers recognized that the program provided them with specific critical techniques for fostering equity in their classrooms. This recognition came about as they realized the value of active learning opportunities to make educational theories meaningful on a personal level. The student teachers also expressed the view that pedagogical activities focusing on issues of equity should begin earlier in the teacher preparation program, rather than at its conclusion.

First-Year Teaching

Several studies of beginning teachers in master's-level teacher education programs indicate that the challenges that prospective teachers experience in providing equitable learning opportunities for diverse student groups continue into their beginning years as classroom teachers. J. A. Bianchini and E. M. Solomon (2003) worked with eight beginning secondary science teachers (five European American, two Asian American, and one Latino) in a course on the nature of science and issues of equity and diversity in a fifth-year teacher education program. The study examined the teachers' views of science and science teaching as they related to issues of equity and diversity along three dimensions: personal, social, and political. The personal dimension involved the use of personal experiences to support students' science learning, and also addressed how to represent science in ways that would be meaningful to all students. The social dimension centered on the question of how to broaden notions of who does science. The political dimension involved science as cultural production. These three dimensions must be present in any science education that aims to be inclusive of all students. Results indicated that beginning teachers routinely drew from their own personal experiences to support their views of the nature of science and find ways to represent science to all students; however, they rarely moved beyond the personal dimension into the social or political dimension.

Bianchini and colleagues (2003) followed three first-year secondary science teachers (all European Americans), recent graduates of a fifth-year teacher education program, into their first year of teaching. The researchers explored these beginning teachers' attempts to present contemporary descriptions of the nature of science and to implement equitable instructional strategies in their classrooms. The results indicate commonalities across the beginning teachers' successes and struggles in learning to teach science in current and equitable ways. The teachers had examined the nature of science, as well as issues of equitable and inclusive science instruction, in their science teacher preparation program, and were able to translate some of what they had learned into their teaching practices. Yet there were other aspects of the nature of science that they had examined in the program but rarely addressed in their teaching: for example, the ways in which social values and cultural biases shape scientists' research questions, methods, and findings, or the kinds of knowledge and practices that indigenous cultures contribute to science. Furthermore, the researchers noted the influence of California's recently adopted state science standards on classroom instruction and teacher learning. The beginning teachers learned from their colleagues that it was crucial that they introduce science content and skills to meet specific standards and to help students excel on standardized tests.

K. Tobin and colleagues (Roth et al., 2004; Tobin, Roth, & Zimmerman, 2001; Tobin et al., 1999) have addressed the difficulties of learning to teach in urban schools characterized by student resistance, violence, lack of instructional resources, high teacher attrition, high student mobility, and inadequate funding. The researchers proposed co-teaching as a model for teacher preparation and the professional development of urban science teachers. The case studies described beginning science teachers during their year-long field experience within a master's program in science education. While teaching science to African American students placed in a low-track program of study in an urban high school, the beginning teachers enacted a curriculum that was culturally relevant to the students and responsive to the students' interests, acknowledged the students' minority status with regard to science, and helped the students meet school district standards in science. In the context of co-teaching, the teachers devised appropriate and timely actions by discussing shared experiences with other educational actors (including peers, cooperating teachers, university supervisors, and high school students). Furthermore, the presence of a co-teacher increased access to social and material resources, and thereby increased opportunities for actions that otherwise would not have occurred.

Preparation of Minority Science Teachers
C. C. Loving and J. E. Marshall (1997) present results of an ongoing evaluation of a funded project designed to recruit, educate, retain, and credential ethnic minority science teachers in a region with large numbers of ELL and non-White students. This project attempted to address the lack of minority role models in science teaching by recruiting college-eligible high school students to major in science and work toward becoming secondary science teachers. There were 13, 15, and 9 students during the first three years of this five-year project. They were African American, Hispanic, Hmong, Cambodian, and Native American, with a balance of male and female students. Multiple methods of qualitative and quantitative evaluation were used, including student and staff questionnaires, course assessment data, instructors' written comments, and interviews with students and key staff members. Key findings included the following:

- The role of the full-time, professional project counselor may be the single biggest factor in the success of students during their first year.

- The project counselor alone cannot serve as sole academic advisor for the various science departments. Ongoing academic advising is critically important.
- Placing teacher mentors and student participants together as learners increased positive attitudes, self-confidence, and content knowledge among students, while also enabling the teacher mentors to better understand nonmainstream students in their own classrooms.
- Assessment of the participants, conducted during the summer, resulted in changes to better meet the needs of the participants in subsequent years.
- Science professors changed the nature of their science course content and teaching strategies as a result of working with project staff.
- Established academic guidelines are important when dealing with students who are not succeeding.
- Students recruited from ethnic minority groups often have special needs, which may or may not be easily recognized.

Teacher Professional Development

Professional development aims to expand and improve the learning opportunities that teachers provide to students by enhancing teachers' knowledge of subject matter and enabling them to provide reform-oriented or standards-based instructional practices (Richardson & Placier, 2001; Wilson & Berne, 1999). Research on professional development has generally focused on the form and structure of professional development efforts, such as total contact hours, whether contact hours are concentrated or distributed, and whether consultation or coaching occurs in the context of in-class visits or entirely outside of the classroom (Desimone et al., 2002; Garet et al., 2001). However, recent literature emphasizes that the substance of professional development activities (i.e., what teachers actually learn) has the greatest impact on teachers' beliefs and practices and, eventually, on student learning outcomes (Cohen & Hill, 2000). According to this literature, the most effective professional development is that which enhances teachers' knowledge of specific subject matter content, their understanding of how children learn that content, and reform-based instructional practices.

Research on professional development indicates that teachers need to engage in reform-oriented practices themselves in order to be able to provide effective science instruction for their students. Teachers need opportunities to develop their own deep and complex understandings of science concepts, recognize how students' misconceptions cause learning difficulties (Kennedy, 1998; Loucks-Horsley et al., 1998), engage in science inquiry themselves to be able to foster student initiative in inquiry (NRC,

2000), and learn how to enable students to negotiate ideas and construct collective meanings about science (Lemke, 1990).

Effecting changes in teachers' knowledge, beliefs, and practices with regard to science instruction is a demanding and arduous process. Teachers who engage in professional development often blend a repertoire of reform-oriented practices with traditional practices (Cohen & Hill, 2000; Knapp, 1997). For example, teachers tend to implement isolated features of reform-oriented practices, such as encouraging students to pose their own questions or using hands-on activities. But they are less likely to help students to make meaning of the data they collect, offer explanations based on evidence, or evaluate their misconceptions. Since few teachers have been sufficiently prepared in terms of their own knowledge of science content and content-specific teaching strategies (Garet et al., 2001; Kennedy, 1998; Loucks-Horsley et al., 1998), reform-oriented practices present challenges for most teachers in science classrooms.

Even within professional development efforts that ostensibly emphasize academic achievement for all students, attempts to link subject matter content to the specific linguistic and cultural experiences of diverse student groups have been limited. Professional development programs focusing on subject matter content generally do not address the ways in which students from diverse backgrounds engage that content, whereas programs focusing on student diversity seldom consider the specific demands of different subject areas. This is especially true with regard to science disciplines, which the field of education has traditionally treated as "culture-free" in terms of both epistemology and pedagogy.

K. King, L. Shumow, and S. Lietz (2001) showed that professional development is critical for elementary teachers in urban schools. Working with four teachers in an urban elementary school with a high percentage of low-income minority children, the researchers found that the teachers were poorly prepared in terms of science content knowledge, instructional skills, and classroom management. Although classroom observations indicated that science lessons were typically expository in nature with little higher-level interaction of significance, the teachers perceived their own teaching practice as hands-on and inquiry based.

Despite the critical need for professional development of science teachers working in diverse classrooms, the literature is extremely limited. A small number of studies, described in this section, examine professional development of science teachers with racial/ethnic minority or low-SES students in inner-city schools and/or urban school districts.

J. Johnson and E. Kean (1992) presented a qualitative study of collaboration between university faculty and science teachers in one school district to improve the learning environments in culturally diverse science classrooms. One-third of the students in the school district were students of color, but only 4 of the 120 secondary science teachers were from nonmainstream backgrounds. The project was based on the "cultural conflict" model for teacher change, which ascribes unequal school outcomes to cultural differences between teachers and students in learning style, cognitive style, interaction style, prior knowledge, and language. During the three-year period, there were 14 participants from the school district (11 science teachers and 3 administrators) the first year, 29 (27 teachers and 2 administrators) the second year, and 12 (all teachers) the third year. All participated in summer workshops and follow-up interactions with university faculty during the academic year. The summer workshops focused on multicultural understanding, problem solving, and cooperative learning. The results indicate positive changes in student–teacher interactions, the role of the teacher, science content, instructional strategies, classroom culture, and relationships with parents and community.

C. Buxton (in press) devised a professional development intervention based on a model of "authentic science inquiry" at an academically low-performing inner-city elementary school. He developed the framework of contextually authentic science inquiry that links the strengths of a canonically authentic model of science inquiry (grounded in the Western scientific canon) with the strengths of a youth-centered model of authenticity (grounded in student-generated inquiry), thus bringing together science content standards and topics with critical social relevance. He applied this framework to an examination of how 14 elementary teachers participating in a master's degree program and their students at this school interpreted and enacted ideas about authenticity and collaboration. Specifically, he investigated the structures in the professional development that were required to support contextually authentic science learning. On the basis of an analysis of 20 one-hour classroom videos and 24 focus group interviews over a two and a half year period, he identified several general principles in the areas of curriculum, instruction, and assessment that frequently led teachers to engage their students in contextually authentic science learning.

Curricular principles included a focus on highly localized neighborhood environments and connections to students' families. Instructional principles included learning to ask testable questions and a focus on how doing science together could help foster communal relationships. Finally, assessment principles included increasing students' choice over how to document their learning, as well as the availability and use of multimedia technologies as assessment tools. Despite pointing to positive examples of contextually authentic science learning that resulted from this professional development intervention, Buxton (2005b) also paints a rather bleak picture of how the professional development was undermined by policies of high-stakes accountability and administrative mandates. The study highlights the challenges of implementing a professional development intervention that is responsive to the needs of inner-city teachers and their students within the current policy context faced by many urban schools.

O. Lee and colleagues (Cuevas et al., 2005; Lambert et al., in press; Lee et al., 2004, 2005) implemented a large-scale instructional intervention aimed at promoting achievement and equity in science, particularly science inquiry, and English language and literacy development for culturally and linguistically diverse elementary students. Grounded on the instructional congruence framework and the teacher-explicit to student-exploratory continuum, the research emphasized the integration of three domains: (a) inquiry-based science instruction, (b) English language and literacy, and (c) students' home language and culture (see "Congruent Instruction" in Chapter 5). The professional development intervention was designed to enhance teachers' knowledge, beliefs, and practices in integrating these three domains for their students across varied classroom contexts. A series of studies (described in the following paragraphs) examined the process and impact of the intervention with participating teachers and the impact of classroom practices on students' science and literacy (writing) outcomes.

Lee and colleagues (2004) focused specifically on professional development efforts in the domain of science instruction. They examined (a) teachers' initial beliefs and practices related to inquiry-based science and (b) the impact of the professional development intervention on teachers' beliefs and practices over the course of the school year. As a schoolwide initiative, the study involved all 3rd- and 4th-grade teachers (53 total) at six elementary schools in a large urban school district. The intervention consisted of provision of two instructional units (plus supplies) at each grade level and four full-day teacher workshops over the course of the school year. In the science instruction domain, the intervention emphasized how to promote scientific understanding, inquiry, and discourse with elementary students from diverse cultural and linguistic backgrounds. Teachers' beliefs were examined by means of a questionnaire and focus group interviews at the beginning and end of the school year, while teachers' instructional practices were examined through one classroom observation in fall and one in spring with each participating teacher. Using both quantitative (dependent t-tests) and qualitative methods, the study reports the results of the first year of implementation of the intervention as part of a longitudinal design. At the end of the school year, teachers reported significantly enhanced knowledge of science content and stronger beliefs about the importance of science instruction, although their classroom practices did not show statistically significant change.

J. Lambert and colleagues (in press) replicated the Lee et al. study (2004) with all 5th-grade teachers (total 23) at the same six elementary schools. At the end of the school year during the first year of the intervention's implementation, the 5th-grade teachers reported significantly enhanced knowledge of science content, as well as teaching of science to promote students' understanding, inquiry, and discourse. (This study did not report teacher practices.)

Beyond examining the impact of the professional development intervention on teachers' beliefs and practices, Lee and colleagues (2005) examined its impact on students. The research addressed three areas of student outcomes: (a) overall science and literacy achievement, (b) achievement gaps among demographic subgroups, and (c) comparison with national (NAEP) and international (TIMSS) samples of students. The research involved 1,523 3rd- and 4th-grade students at the six elementary schools. Significance tests of mean scores between pre- and posttests indicated statistically significant increases on all measures of science and literacy at each grade level. First, on paper-and-pencil science tests (including multiple-choice, short-answer, and extended response items), both 3rd- and 4th-grade students showed statistically significant gains and large effect magnitudes at the end of the school year. Students showed similar results on writing prompts as measures of literacy development. Second, although at the beginning of the school year the students performed lower than 3rd/4th-grade national and international samples of students on NAEP and TIMSS items (mostly multiple-choice items with a small number of short-answer and extended response items), they generally performed higher than 3rd/4th-grade samples, and comparable to or higher than 7th/8th-grade samples, at the end of the school year. Third, achievement gaps narrowed significantly on some of these measures among demographic subgroups (defined in terms of ethnicity, home language, ESOL level, SES, special education status, and gender). Finally, as students advanced from one grade level to the next, the intervention seemed to have cumulative effects on achievement gains and narrowing of achievement gaps.

P. Cuevas, and colleagues (2005) examined the impact of the intervention on (a) children's ability to conduct science inquiry overall and to use specific skills in inquiry, and (b) narrowing the gaps in children's ability to conduct science inquiry among demographic subgroups of students. The study involved 25 3rd- and 4th-grade students taught by seven teachers who were selected for their effectiveness in teaching science and literacy to students of diverse linguistic and cultural backgrounds. The teachers selected these students to represent different achievement levels (high and low) and gender groups. Since the students came from all six schools, they also represented different ethnicities, SES levels, home languages, and levels of English proficiency. At the beginning and end of the school year, the students participated in elicitation sessions in which they were asked individually to design an experiment regarding the effect of surface areas on the rate of evaporation. Results of paired samples of t-tests indicate that the intervention enhanced the inquiry ability of all students, regardless of demographic background. Particularly, low-achieving, low-SES, and ESOL-exited students made impressive gains.

The results of the studies by Lee and colleagues indicate teachers' overall receptiveness to the intervention, as well as its relative strengths and weaknesses with regard to the professional development goals. The results also indicate the positive impact of the intervention on students' science and literacy achievement and on narrowing of achievement gaps among demographic subgroups. Given that the research included all 3rd-, 4th-, and 5th-grade teachers within the six participating schools, rather than a self-selected group of volunteer teachers with an interest in "teaching science for diversity," their beliefs and practices may be more representative of teachers in general. Thus, the results have implications for further large-scale implementation (i.e., scaling-up) of the intervention with diverse student groups in urban school districts.

J. B. Kahle and colleagues conducted a series of studies to examine the impact of standards-based teaching practices (i.e., extended inquiry, problem solving, open-ended questioning, and cooperative learning) on the science achievement and attitudes/perceptions of urban African American middle school students (Boone & Kahle, 1998; Damnjanovic, 1998; Kahle, Meece, & Scantlebury, 2000). These studies were part of the NSF-supported Ohio Statewide Systemic Initiative (SSI) known as "Discovery." This reform initiative was grounded in sustained professional development for middle school science and mathematics teachers. The Ohio SSI's professional development programs consisted of six-week summer institutes in physics, life science, and mathematics, and six full-day seminars during the academic year.

In the Kahle et al. study (2000), a random sample of 126 schools was drawn from all schools in the state of Ohio that enrolled students from grades 5 through 9 and that had at least one teacher who participated in the Ohio SSI's professional development programs. Then, a subsample of the eight schools was identified on the basis of at least 30% minority student enrollment. At each of these eight schools, one teacher was randomly selected from among those who had completed the SSI's professional development program. Each SSI teacher was then matched with a non-SSI teacher in the same school who taught a similar class. The non-SSI teachers volunteered to administer student achievement and attitude measures and to complete the same teacher questionnaire that the SSI teachers completed. When the data were reviewed, 18 teachers (8 SSI and 10 non-SSI) reported data for their students. Fifteen were White (6 SSI teachers and 9 non-SSI teachers), and three were African American (2 SSI teachers and 1 non-SSI teacher).

The student sample consisted of all of the African American 7th- and 8th-grade students enrolled in the SSI and non-SSI teachers' classes. Only 7th- and 8th-grade students were involved because the achievement tests were composed of NAEP public items (age 13+). Students' science achievement was measured using a total of 29 NAEP public release items, of which 20 involved solving problems or conducting science inquiry.

Additionally, students responded to questionnaires focusing on: (a) their attitudes toward science, (b) their teachers' use of standards-based science teaching practices, (c) their parents' involvement in science homework and projects, and (d) their peers' participation in science activities. Hierarchical linear modeling (HLM) procedures were used to examine the predictive influence of student and teacher variables on science achievement and attitudes. Results indicate that students of SSI teachers rated their teachers as using standards-based teaching practices more frequently than did students in non-SSI teachers' classes. Additionally, compared to students of non-SSI teachers, students of SSI teachers scored higher on the science achievement test and had more positive attitudes toward science. The improvements were most pronounced with African American boys. These results suggest that professional development designed to enhance teachers' content knowledge and use of standards-based teaching practices can not only improve science achievement overall but also reduce inequities in achievement patterns for urban African American students.

A. Damnjanovic (1998) analyzed the impact of the Ohio SSI's professional development programs on science achievement by race and gender. The study used an ex post facto research design, and was based on ANOVA and multiple regressions of quantitative data, as well as qualitative methods using observation and interview data. A total of 610 7th- and 8th-grade students enrolled in the SSI teachers' classes participated in the study, including 190 African American females, 131 White females, 168 African American males, and 121 White males. Three sets of results are reported. First, on the NAEP public release items (described in the previous paragraph), females scored significantly higher on the science achievement test than males, and White students of both sexes scored higher than African American students, but there was no interaction of race and gender. Second, for the student sample as a whole, contemporary classroom teaching (i.e., science inquiry and cooperative group work), students' positive attitudes toward science, and students' participation in hands-on/problem-solving activities were significant positive predictors of science achievement. In contrast, low peer interest and involvement in science was a significant negative predictor of science achievement. Finally, predictors for science achievement varied for each race and gender group.

W. J. Boone and Kahle (1998) analyzed the impact of the Ohio SSI's professional development programs on students' perceptions of science, disaggregating the data by race and gender. In total, more than 900 middle school science students completed the questionnaires in 1995 and 1996 (but no ethnic breakdown is reported in the article). Chi-square tests were used to compare student responses between the SSI teachers' and non-SSI teachers' students. Responses of both African American and White students, boys as well as girls, indicate that all students actively participated in those science classes taught by SSI teachers. With the exception of White boys, students were more involved in active learning in SSI classes compared to non-SSI classes.

On the basis these results, Kahle and colleagues (Boone & Kahle, 1998; Damnjanovic, 1998; Kahle, Meece, & Scantlebury, 2000) offer policy recommendations to foster the use of inquiry-based science instruction with urban African American students at state and district levels, with regard to length of instructional period, availability of appropriate materials and supplies, and use of performance-based and other types of authentic assessments. They also offer policy suggestions to facilitate teachers' participation in sustained professional development that enhances both content knowledge and skills in using standards-based teaching strategies. Such policies should address support and/or incentives for teacher participants, appropriate mechanisms for evaluating teachers who use standards-based instruction, possible negative effects of traditional norm-referenced tests on the use of inquiry teaching, and provision of professional development activities that take into account teachers' access to the time and place of such activities.

Teacher Education with ELL Students

Teachers of ELL students need to promote students' English language and literacy development as well as academic achievement in subject areas. This may require subject-specific instructional strategies that go beyond the general preparation in English to Speakers of Other Languages (ESOL) or bilingual education that many teachers receive. Unfortunately, a majority of teachers working with ELL students believe they are not adequately prepared to meet their students' learning needs, particularly in academically demanding subjects such as science, mathematics, and reading (National Center for Education Statistics, 1999). Most teachers also assume that ELL students must acquire English before learning subject matter, though this approach almost inevitably leads such students to fall behind their English-speaking peers (August & Hakuta, 1997; García, 1999).

Professional development to promote science as well as English language and literacy development with ELL students involves teacher knowledge and practices in multiple areas. First, in addition to ensuring that ELL students acquire the language skills necessary for social communication, teachers need to promote ELL students' development of general and content-specific academic language functions, such as describing, explaining, comparing, and concluding (Wong-Fillmore & Snow, 2002). Second, teachers must be able to view language within a human development perspective. Such an understanding enables them to formulate developmentally appropriate expectations about language comprehension and production over the course of students' learning of English. Finally, teachers need to be able to apply this knowledge to the teaching of general and content-specific academic language. The amalgamation of these three knowledge sources should result in teaching practices that (a) engage students of all levels of English proficiency in academic language learning, (b) engage students in learning activities that have multiple points of entry for students of differing levels of English proficiency, (c) provide multiple modes for students to display learning, and (d) ensure that students participate in a manner that allows for maximum language development at their own level.

Fradd and Lee (1995) examined teachers' perceptions of science instruction at two elementary schools, one suburban and one urban, with high percentages of ELL students. Their study was conducted through formal and informal interviews with teachers. The results indicated that teachers in both schools viewed science instruction positively, expressed beliefs that all students could learn science, and stressed that science learning opportunities should be available to all students. They also agreed on the importance of active student engagement, practical applications in daily life, and authentic and meaningful tasks. They emphasized the need to promote language development during science instruction for all students. Despite these similarities, the two schools displayed clear contrasts in terms of teachers' ideas about opportunities and resources for science learning and the instructional environment in each school setting. The urban schoolteachers perceived students' limited English proficiency and cultural difference as reasons for their difficulties in learning science. The teachers were not specific about instruction or articulate about their own beliefs regarding effective instructional approaches. In contrast, the suburban schoolteachers generally promoted science learning along with English language skills more effectively than those at the urban school (although it should be recognized that ELL students at the suburban school were likely to have better academic skills in the home language than those at the urban school). However, even under these more favorable conditions, the suburban teachers missed opportunities to promote student learning, as their science instruction tended to involve discrete science activities rather than being organized around a comprehensive science program.

A limited number of studies, described as follows, address professional development efforts to help practicing teachers expand their beliefs and practices in integrating science with literacy development for ELL students. These studies range from large-scale professional development at the school or district level to intensive teacher research with small numbers of participants.

T. Stoddart and colleagues (2002) recognize that it is often not possible to teach academic subjects to ELL students in their native language while they are acquiring proficiency in English. A chronic shortage of bilingual teachers, particularly teachers qualified to teach science, means that few ELL students receive content instruction in their primary language. Additionally,

English-only legislation in an increasing number of states severely limits the teaching of academic subjects in languages other than English. As an alternative, the researchers attempted to integrate the teaching of academic subjects, such as science, with second language (i.e., English) development. The thesis of this research is that inquiry-based science provides a particularly powerful instructional context for the integration of science content and second-language development with ELL students.

As part of an NSF-supported local systemic initiative, the Stoddart et al. study involved 24 elementary schoolteachers of predominantly Latino ELL students. The researchers developed a five-level rubric to assess teachers' understanding of science and ESOL integration that was based on a conceptual framework for integrating English language development with inquiry-based science. Then, on the basis of interviews with the 24 teachers, they provided exemplars of teacher thinking at each level in the rubric. The preliminary analyses of teachers' work during the five-week summer professional development program indicate changes in teachers' understanding of science and language integration. Prior to their participation, the majority of teachers viewed themselves as well prepared to teach either science or language, but not both. After their participation in the professional development program, the majority of teachers believed they had improved in the domain in which they had initially felt less prepared. This change typically involved a shift from a restricted view of the connections between inquiry science instruction and second-language development to a more elaborated reasoning about the different ways that the two could be integrated.

As part of the ongoing research described earlier, Lee (2004) examined patterns of change in elementary teachers' beliefs and practices as they learned to teach English language and literacy as part of science instruction throughout their three-year collaboration with the research team. Working with six bilingual Hispanic teachers of 4th-grade Hispanic students at two elementary schools, Lee described changes in teachers' beliefs and practices around literacy instruction, which were initially broad and general but gradually came to a focus on specific aspects of English language and literacy in the context of science instruction. Teachers came to adapt their literacy instruction and provide linguistic scaffolding to meet students' learning needs. They also helped students to acquire the conventions of standard oral and written English, including syntax, spelling, and punctuation, in social and academic contexts. Additionally, they used multiple representational formats in oral and written communication to promote both literacy and science learning. Overall, science instruction provided a meaningful context for English language and literacy development, while language processes provided the medium for understanding science.

As an expansion of the Lee (2004) study, J. Hart and Lee (2003) provided similar professional development opportunities to all 3rd- and 4th-grade teachers (53 total) from six elementary schools serving students from a range of ethnic, linguistic, and SES backgrounds and levels of English proficiency. As part of a longitudinal professional development intervention, their study focused specifically on assisting teachers in integrating English language and literacy development as part of science instruction with ELL students. It examined teachers' initial beliefs and practices, and the extent of change in teachers' beliefs and practices, following the first year of implementation of the intervention. Through the provision of curricular materials and teacher workshops, the intervention focused on how to incorporate reading and writing into science instruction, integrate appropriate (English) grammar into science instruction, and provide linguistic scaffolding to enhance science meaning. Data sources included teacher questionnaires and focus group interviews at the beginning and end of the school year and classroom observations in fall and spring. Both quantitative (using dependent t-tests) and qualitative results indicate positive change in teachers' beliefs and practices, as teachers came to place greater emphasis on the importance of reading and writing in science instruction, express a broader and more integrated conceptualization of literacy in science, and provide more effective linguistic scaffolding to enhance scientific understanding.

As a result of the instructional intervention, 3rd- and 4th-grade ESOL students in the study showed statistically significant gains in science and literacy (writing) achievement at the end of the school year (see the description of Lee et al., 2005, in "Teacher Professional Development" in Chapter 7). They also demonstrated enhanced abilities to conduct science inquiry (see the description of Cuevas et al., 2005). Especially at the end of the school year, bilingual Spanish/English-speaking students and those who exited from ESOL programs showed science and literacy achievement scores that were comparable to or higher than those of monolingual English-speaking students, thus narrowing achievement gaps.

Amaral and colleagues (2002) examined professional development in promoting science and literacy with predominantly Spanish-speaking elementary students as part of a district-wide local systemic reform initiative. The inquiry-based science program started with 14 pioneer, volunteer teachers from two school sites. As the program progressed, more teachers and sites were added to the program until the program became available to all teachers at all elementary schools in the school district. Over four years, teachers were provided with at least 100 hours of professional development designed to deepen their understanding of science, address pedagogical issues, and prepare them to teach science at their grade level. Teachers also received in-classroom professional support from a cadre of resource teachers, and complete materials and supplies for all the science units. Results indicate that the science and (English) literacy achievement of ELL students increased in direct relation to the number of years they participated in the program (see "English Language and Literacy in Science Instruction in Chapter 5). English-proficient students performed significantly better than limited English-proficient students in both science and writing.

As part of their ongoing Chèche Konnen Project, Warren and Rosebery (1995) adopted a sociocultural view of teaching and learning in their description of how teachers practiced science as members of a scientific community. The researchers organized a seminar on scientific sense making and worked with eight teachers, including five bilingual education teachers, two ESL teachers, and a science specialist. The teachers and the research team met every other week for two hours after school during the school year and for two weeks in the summer. They engaged in doing science as well as thinking about science as a discourse with particular sense-making practices, values, beliefs, concepts, objects, and ways of interacting, talking, reading, and writing. As they conducted scientific investigations around their own questions and shared their work with colleagues, the teachers learned to appropriate the discourse of science. They also experienced success in creating classroom communities in which students' scientific questions were valued, while they continued to reflect on ways to help shape students' questions into scientific investigations.

Ballenger and Rosebery (2003) explored a particular approach to teacher research, based in teachers' concerns for underachieving students, especially those from nonmainstream backgrounds. They reported on a conference where experienced teachers from existing teacher research groups met with new teachers to explore classroom data together. The conference was structured around joint exploration of children's classroom talk and work, with special attention to the talk and work of "puzzling children," that is, those a teacher found difficult to understand. The experienced teacher researchers showed how close observation of children could challenge taken-for-granted assumptions about children's talk and work. They also demonstrated that children who made puzzling responses did not necessarily have deficient ideas, but rather were operating from a framework different from the one commonly assumed.

Discussion

The different areas of research on science teacher education for nonmainstream students present several notable features. First, the studies on science teacher preparation all employed case study or other qualitative approaches to examine science methods courses, student teaching, and first-year teaching in teacher preparation programs. They were conducted by researchers who were either instructors for science methods courses or supervisors for student teaching. In other words, the researchers conducted research on their own practices and students, from the perspective of both science education researchers and teachers who were committed to teaching science for student diversity. In most cases, prospective science teachers were from mainstream backgrounds (with the exception of Brand & Glasson, 2004), and for the first time engaged issues of student diversity in a university course or during their student teaching. All the studies examined the prospective teachers' beliefs and practices as they interacted with course instructors and peers in a university setting or with students in a classroom setting.

Second, the small number of studies on teacher professional development involved teachers across schools in a school district or a state (Boone & Kahle, 1998; Cuevas et al., 2005; Damnjanovic, 1998; Johnson & Kean, 1992; Kahle et al., 2000; Lambert et al., in press; Lee et al., 2004; Lee et al., 2005). These studies examined different aspects of the impact of longitudinal interventions – Johnson and Kean focusing on instructional practices, Lambert et al. and Lee et al. focusing on change in teachers' beliefs and practices; Cuevas et al. and Lee et al., focusing on students' science and literacy achievement, and Kahle and colleagues focusing on students' science achievement and attitudes/perceptions. The studies employed different research methods: Johnson and Kean using descriptive methods to gather qualitative data in classroom settings; Lambert et al. and Lee et al. using questionnaires, interviews, and classroom observations to gather both quantitative and qualitative data with teachers; Lee et al. using both project-developed and standardized tests to gather quantitative data about students' science and literacy achievement; Cuevas et al. using a science inquiry task to gather qualitative and quantitative data about students' inquiry abilities; and Kahle and colleagues using questionnaires and standardized tests to gather quantitative data about students' science achievement and attitudes/perceptions.

Finally, the small number of studies on teacher education with ELL students involved professional development interventions, with no study involving prospective science teachers in teacher preparation programs. These studies used a range of research methods, including ethnographic research with a small number of participants (Warren & Rosebery, 1995) to large-scale research on school- or district-wide initiatives (Amaral et al., 2002; Hart & Lee, 2003). Some focused on the integration of science and English language and literacy (Hart & Lee, 2003; Lee, 2004; Stoddart et al., 2002), while others focused exclusively on science instruction. Amaral et al. (2002), Cuevas et al. (2005), and Lee et al. (2005) examined the impact of the professional development intervention on students' science and/or literacy achievement, whereas the others examined the impact on teachers' beliefs or practices. Notably, none of them portrayed teacher education programs that posited a significant instructional role for students' home languages other than English, or that encouraged teachers to explore how ELL students' abilities in the home language might support their development of scientific understandings. Instead, they stressed the teacher's role in terms of promoting ELL students' acquisition of English in the context of science instruction.

Across the three areas of the research, the results indicate a range of overall patterns in teachers' beliefs and/or practices after their participation in professional development activities: (a) teachers who were already committed to embracing student diversity in science education and became more committed through professional development opportunities; (b) teachers who had not considered student diversity but came to recognize and accept it as important in science education; (c) teachers who remained unconvinced of the importance of student diversity in science education; and (d) teachers who actively resisted embracing student diversity in general, and in science education in particular. Additionally, even those teachers who came from racial/ethnic minority backgrounds (Bland & Glasson, 2004) and those who were committed to educational equity (e.g., Luft, 1999; Luft et al., 1999) still faced challenges related to student diversity in their teaching. Even when changes in teacher beliefs and practices occurred, such changes were demanding and slow.

A few studies involved all teachers in a given grade or from entire schools, rather than volunteer teachers (Amaral et al., 2002; Hart & Lee, 2003; Lambert et al., in press; Lee et al., 2004). Research on schoolwide professional development initiatives reveals both advantages and limitations of such programs (Blumenfeld et al., 2000; Fishman et al., 2004; Gamoran et al., 2003; Garet et al., 2001). On the one hand, collective participation of all teachers from the same school or grade level in professional development activities allows teachers to develop common goals, share instructional materials or assessment tools, and exchange ideas and experiences arising from a common context. On the other hand, unlike programs comprised of volunteer teachers seeking opportunities for professional growth, programs that are implemented schoolwide inevitably include teachers who are not interested in or even resist participation. Additionally, the intensity of professional development activities may be compromised due to limits on the number of days teachers may be out of their classrooms, the difficulty of finding large numbers of substitute teachers, the pressure to prepare students for high-stakes assessment, or other such constraints. Despite these hurdles, schoolwide professional development can provide valuable insights for large-scale implementation.

8

School Organization and Educational Policies

Classroom practices occur in the context of school policies and institutional structures, which are shaped by policies mandated by the individual school district, the state, and the nation. Policies are interpreted and mediated by educational actors at every level of their implementation, to the extent that they are sometimes implemented in ways that are directly contrary to their presumed goals. A limited body of literature highlights features of school organization that influence science teaching for students from nonmainstream backgrounds. Another limited body of literature describes how district, state, and federal policies on science instruction and assessment influence classroom practices and science learning.

School Organization

The limited literature on school organization in relation to science education for diverse student populations addresses such issues as tracking, school restructuring, school leadership, and resources to promote change in instructional practices and student learning. In general, these factors affect the learning opportunities available to nonmainstream students more than those available to mainstream students, since the latter more often enjoy other supports for their science learning (e.g., better-equipped schools, highly educated parents, etc.). In contrast, the academic success of nonmainstream students depends more heavily on an adequate school environment, and it is precisely these students who are less likely to have access to such environments, particularly in inner-city schools. As described in this chapter all the studies focus on urban contexts.

Tracking
Tracking or ability grouping results in differential learning opportunities for different groups of students (Oakes, 1990). In theory, such practices

separate the academically stronger from the academically weaker students; in practice, this often means separating students of wealthier families from those of less wealthy families, and mainstream students from nonmainstream students. Regardless of initial achievement levels, students who are placed in low-track courses demonstrate lesser gains over time than those placed in higher-level courses. Disadvantaged students are often placed in low-track courses that require fewer high-level cognitive activities, provide less challenging academic content, and generally represent lower expectations for those students. Educational scholars are in general agreement that tracking or ability grouping creates a cycle of restricted opportunities, diminished outcomes, and exacerbated inequalities for students from poor and nonmainstream backgrounds; nevertheless, it remains a common practice in schools throughout the nation.

School Restructuring

School restructuring efforts (which often address tracking, among other issues) can narrow SES- and ethnicity-based achievement gaps. V. Lee and colleagues (Lee & Smith, 1993, 1995; Lee et al., 1997) conducted a series of studies focused on mathematics and science education to examine how the structure of high schools affects both overall student learning and equitable distribution of learning by SES.

Lee and Smith (1995) examined how practices consistent with the restructuring movement influenced the learning and academic engagement of high school students in a large and nationally representative sample of schools. The study focused on 30 practices that either were or were not being implemented in the schools, according to reports from the schools' principals. Of these practices, 12 were classified as being consistent with the restructuring movement; the other 18 were considered to be more representative of traditional educational experiences. The study used hierarchical linear modeling (HLM) to examine the impact of these practices on student learning. The results indicate that student achievement was higher in schools that reported 3 or more reform-oriented practices than in those that reported 2 or fewer, but which 3 did not seem to matter. In those schools that engaged in practices consistent with the restructuring movement, achievement and engagement were significantly higher (i.e., schools were more effective), and differences in achievement and engagement among students from different SES backgrounds were reduced (i.e., schools were more equitable).

V. Lee and colleagues (1997) examined the link between broader organizational practices and student learning in mathematics and science in high schools. Their study examined whether differences in the social and academic organization of high schools would help explain the positive effects reported in Lee and Smith (1995). It used nationally representative and longitudinal data from the NELS:88 and employed HLM models to analyze multiple test scores nested in students who are, in turn, nested in schools. The results indicate that academic achievement is influenced by individual practices consistent with the restructuring movement (replicating the results in Lee & Smith, 1995), but is influenced even more by broader organizational attributes that reflect the willingness of school personnel to adopt and adhere to policies and practices leading to high achievement. Specifically, more effective and equitable schools created smaller organizational units ("schools within schools") within the walls of large high schools. They resembled communities rather than bureaucracies, and teachers had strong professional communities that focused on the quality of instruction. They had a strong academic focus, and all students took a highly academic curriculum with limited tracking options.

School Leadership and Resource Use
Several studies examined how school leaders (administrators, teachers, and staff) acquire and use resources (human capital, social capital, and materials) to promote change in science education. J. P. Spillane and colleagues (2001) examined how leaders at one urban elementary school successfully brought together resources to enhance science instruction in a context in which other subjects (i.e., reading, writing, and mathematics) commanded the bulk of the resources by virtue of tradition and formal policy. The researchers argue that promoting change in science education involves: (a) the identification and activation of material or physical resources (i.e., time, money, and other material assets); (b) the development of teachers' and school leaders' human capital (i.e., the individual knowledge, skills, and expertise that form the stock of resources available in an organization); and (c) the development and use of social capital (i.e., the relations among individuals in a group or organization, and such norms as trust, collaboration, and a sense of obligation). The identification and activation of material resources, the development of teachers' and school leaders' human capital, and the recognition and use of social capital inside and outside the school must all be managed simultaneously with an eye toward accountability measures.

The researchers emphasize the importance of "distributed leadership" for bringing about instructional change. Distributed leadership differs from individual leadership (typified by the school principal). Through distributed leadership, different formal and informal leaders bring their different knowledge and skills (i.e., human capital) to the task of promoting change. Additionally, individuals who trust one another (i.e., who share social capital) are more likely to pool their knowledge and skills to promote change. Furthermore, social capital in the form of networks is essential if schools are to tap essential resources such as time and materials (i.e., material resources) in their environment.

A. Gamoran and associates (2003) examined how teachers from elementary through high school in six school districts across the nation taught mathematics and science with diverse student groups. The research involved six case studies of schools that had participated in "design collaboratives," in which research teams, in collaboration with teachers and administrators, designed classroom environments to foster student understanding of subject matter. Within this overall goal, the six schools varied in terms of community demographics, reform context, school organization, number of participating teachers, and relationships with research teams. Additionally, research foci among the six schools varied in terms of subject (mathematics or science) and grade level. Three of the six schools, including one elementary, one middle, and one high school (the three urban schools; the other three were suburban schools) had high proportions of nonmainstream students. The research used observations, interviews, and questionnaires with teachers and district and school administrators over a five-year period.

The results indicate that teachers at the three urban schools had varying conceptions of student diversity and equity. Such conceptions involved ability grouping (tracking) of students; individual differences in learning styles; attributes of group memberships such as SES, race/ethnicity, culture, and language; differential access to curriculum in response to increased accountability measures; and structural inequalities regarding learning opportunities by virtue of group memberships among students. Furthermore, teachers' conceptions of diversity and equity were associated with the acquisition and use of resources (human, social, and material). Successful efforts entailed the strategic use of resources by school personnel to promote change among teachers, and eventually to enable students to learn mathematics and science. The challenges involved in such strategic use of resources were more formidable in urban schools. Since resources and funding tended to be scarcer in urban schools, provision of human, social, and material resources was critical to the support of teacher growth and reform-oriented classroom practices. These researchers also advocate "distributed leadership" (described earlier by Spillane et al. 2001) among school administrators, teachers, and staff to support and sustain the professional community and to bring about change in school policies and practices.

M. S. Knapp and M. L. Plecki (2001) provide a conceptual framework for understanding what is involved in renewing urban science teaching. A central feature is the allocation and actualization of fiscal, temporal, human, and material resources for instructional practices and student learning. At the classroom level, teachers and students should invest and actualize their available resources, including instructional time, intellectual resources that teachers and students bring to their work, and social resources residing in teachers' and students' attitudes toward science learning and each other.

A school's capacity for improving science education includes time as a schoolwide resource (e.g., allocation of time for teacher professional development) and schoolwide intellectual and material resources (e.g., expertise of professional developers or subject area specialists, money to cover substitute teachers' time). The researchers argue that attempts to improve science teaching, or any subject for that matter, can be thought of as a process of strategically "investing" these resources, while establishing such conditions that favor a "return" on this investment.

Educational Policies

All the studies in the limited literature on policies addressing student diversity in science education focus on urban contexts. These studies examine three issues: (a) systemic reform, (b) scaling-up of educational innovations, and (c) accountability. Achieving equity in urban schools is particularly urgent, considering that large urban districts educate 25% of all K–12 students, 30% of all poor students, 30% of all ELL students, and nearly 50% of all ethnic minority students (Pew Charitable Trust, 1998). Although educational policies influence all districts and all schools, their consequences are especially critical in urban schools because of the sheer number of students attending urban schools, the array and scope of the obstacles facing these schools, and the institutional precariousness under which they operate.

Systemic Reform

A set of studies has examined systemic reform to improve science education in urban schools. Systemic reform involves restructuring various components of an education system (generally at the state or district level) in dynamic, interactive, and coherent ways to improve the quality of the curriculum and instruction delivered to all students (Smith & O'Day, 1991). Among the chief concerns is the potential impact of systemic reform on educational equity, for example, how such a system deals with linguistic, cultural, and socioeconomic diversity and how nonmainstream students fare under such a system (McLaughlin et al., 1995; Porter, 1995).

Grounded in the claim that educational reform must be both systemic and equitable, J. B. Kahle (1998) developed an "equity metric" to monitor the progress of reform toward or away from educational equity over time. On the basis of an analysis of NELS: 88, High School and Beyond, and TIMSS, the researcher proposes indicators of equity that are applicable for states, districts, schools, and classrooms. Key indicators are grouped into three categories that are seen as critical for equitable education: *access, retention*, and *achievement* of all students in high-quality science and mathematics programs. Each state, district, school, or class that aims to meet the needs of all students equitably needs to select and use the indicators that are most appropriate for its situation. As progress is made with some groups, other groups of students may suffer setbacks, so that the balance of different indicators and criteria must be continually monitored and adjusted.

The equity metric developed by Kahle (1998) was applied to assess the progress toward equitable systemic reform in several school districts and individual schools (Hewson et al., 2001; Kahle & Kelly, 2001; Rodriguez, 2001). Using the equity metric, P. W. Hewson and colleagues (2001) assessed the progress toward equity of two urban middle schools engaged in the systemic reform of science education. The researchers provide case studies of these two schools, which were part of the Ohio SSI. The equity metric provided an analytical framework for mapping each school's readiness for and progress toward reform, as well as barriers impeding and factors facilitating reform. The results indicated that the culture and climate of the two schools differentially affected their progress toward equitable science education reform. At one school, a combination of factors consumed the attention of science teachers, leaving them with little time or energy to teach science. At the other school, in contrast, science teachers worked in a cooperative, stable environment that provided the time and space to focus their energies on teaching science. The researchers conclude that equitable systemic reform of science education in urban schools requires cohesion between the school and community around clearly understood and accepted goals; responsible and accessible leadership; teachers who feel efficacious, autonomous, and respected; and a community that is supportive and involved. A hallmark of a school that has instituted equitable systemic reform is the school's visible progress toward goals that are clearly understood and accepted by all major stakeholders and that are consistent with state and district criteria.

A. J. Rodriguez (2001) extends Kahle's notion of an equity metric to move beyond "gap gazing" and identify strategies that have the potential to affect issues of access, retention, and achievement of traditionally underserved urban students in science and mathematics. Rodriguez proposes a conceptual framework that incorporates ideological, pedagogical, and operational approaches to systemic reform. With this framework as a foundation, he argues that systemic reform should involve the following key components at four different levels: (a) policy, curriculum, assessment; (b) growth in student achievement and participation; (c) professional development to change the culture of teaching and encourage community support and participation; and finally (d) strategies for scaling-up and for making systemic change self-sustaining.

Rodriguez then applied this framework in order to explore how these key components of systemic reform were implemented and how progress toward equity could be measured in a school district. The case study was conducted in a large urban school district with support from the NSF

Urban Systemic Initiative (NSF-USI), which was aimed at improving the science and mathematics achievement of all students in the district. The results illustrate how all the different components in the conceptual framework are interdependent and necessary for moving reform efforts forward. Rodriguez concludes that the goal of systemic reform should not be to follow a one-model-fits-all approach. Since the complexity of variables at play in large education systems is monumental and always changing, the goal cannot be the control of all variables. Instead, insights generated from promising reform initiatives can stimulate other school districts to design more equitable and inclusive systemic reform efforts tailored to their own contexts.

Since 1993, the NSF-USI had been a catalyst for large-scale systemic change aiming to improve the science and mathematics achievement of all students in 22 large urban school districts. The NSF's "Six Drivers of Systemic Reform" provide a framework for USI implementation. Four "process drivers" include (a) standards-based curriculum, instruction, and assessment; (b) coherent and consistent policies that support high quality learning and teaching; (c) convergence of educational resources; and (d) broad-based support from partnerships and leadership in local communities. The remaining two "outcome drivers" are (a) measures of effectiveness focused on student outcomes and (b) achievement of all students, including those historically underserved.

J. J. Kim and colleagues (2001a) examined the impact of NSF-USI on teacher professional development, curriculum and subject content, and teaching practices at eight NSF-USI sites in 1999 and 2000. The results from surveys of a sample of elementary and middle school teachers indicate that 80%–90% of NSF-USI teachers were actively involved in professional development, which focused on content standards, in-depth study of content, curriculum implementation, multiple strategies for assessment, and new methods of teaching. Teachers reported that as a result of professional development, they were using and applying new methods and standards in their classrooms, with an average of 2.3 on a scale of 0 to 3 (0 = no emphasis, 1 = slight emphasis that accounts for less than 25% of the time, 2 = moderate emphasis that accounts for 25% to 33% of the time, and 3 = sustained emphasis that accounts for more than 33% of the time). State and district curriculum frameworks or content standards, as well as national content standards, had the greatest positive influence on teaching practices. Teachers with more professional development in standards-based curriculum and instruction reported employment of teaching practices that were more consistent with state and national content standards than those employed by teachers with less professional development.

Kim and colleagues (2001b) presented preliminary findings of all 22 NSF-USI programs. The results indicated gains in student achievement, with the greatest gains seen in school districts with the longest period of participation in the NSF-USI programs. Students made overall gains in science and mathematics, and achievement gaps among racial/ethnic groups decreased. Additionally, students substantially increased their enrollment and completion rates in gatekeeping and higher-level mathematics and science courses. Underrepresented minority students made even greater enrollment gains than their peers during the same period, resulting in reduced enrollment disparities. The increasing numbers of 11th- and 12th-grade students taking college entrance examinations (AP, SAT, and ACT) indicated that more students had aspirations of pursuing postsecondary education.

These advances were accompanied by evidence that urban districts had developed the infrastructure to sustain achievement gains. NSF-USI programs established policies that encouraged enrollment and completion in gatekeeping and higher-level mathematics and science courses (e.g., elimination of tracking, increased requirements for high school graduation, inclusion of ESOL and special education students in regular classrooms, and safety-net programs for students experiencing difficulty with challenging mathematics and science courses). NSF-USI programs invested heavily in professional development, believing it to be a key lever for improving student outcomes (see the results in Kim et al., 2001b, described above). The infrastructure of partnerships at each site provided strong support for systemic change based on the site's unique capacities, resources, and experiences. The partners and school districts interacted as part of a unified effort to promote and support large-scale reform. Finally, NSF-USI programs promoted data-driven accountability systems.

Scaling-Up of Educational Innovations

While systemic reform efforts in mathematics and science date from the 1990s to the present (supported largely by the NSF), strategies for the scaling-up of educational innovations have emerged more recently, driven by the current policy context of NCLB Act requirements. Scaling-up involves implementing specific educational innovations or interventions on a large scale, for example, across an entire school district or in numerous schools scattered throughout the nation. In the climate of standards-based reform and accountability, scaling-up is increasingly called for to bring about system-wide improvements (Elmore, 1996). Despite the urgent need for such improvements, the conceptualization of scaling-up is still under development. For example, C. E. Coburn (2003) challenges traditional definitions that equate scaling-up with increasing the number of teachers, schools, or districts involved. Instead, she proposes a conceptualization that addresses "the complex challenges of reaching out broadly while simultaneously cultivating the depth of change necessary to support and sustain consequential change" (p. 3).

Stages of going to scale are also under development (Raudenbush, 2003). The first stage involves designing an innovation, whether a whole-school reform program, a professional development initiative, curriculum materials, instructional strategies, computer technology, or some combination of these. Once an innovation is designed, the next stage involves testing its efficacy – the magnitude of the treatment effect under extremely favorable conditions. "Efficacy studies" are important for demonstrating that a new treatment can have a significant impact on desired outcomes (i.e., to test the theory underlying the innovation). Once efficacy studies show that an innovation works, the next stage involves testing its ability to go to scale. "Effectiveness studies" examine the magnitude of the treatment effect under the constraints of field settings (i.e., to test implementation of the innovation). Once effectiveness studies demonstrate a positive impact, an innovation can be taken to scale across districts and even states, without the involvement of the original designers.

Scaling-up efforts occur within the demands and constraints of educational policies, local institutional conditions, limited resources, individual teacher practices, expectations of local stakeholders, and other factors. Scaling-up inevitably compromises conceptual rigor and fidelity of implementation to some degree, by subjecting the proposed innovation to the realities of varied educational contexts. Furthermore, scaling-up in multilingual, multicultural, or urban contexts involves numerous challenges, due to fundamental conflicts and inconsistencies in defining what constitutes effective educational policy and practice, as well as inequitable distribution of educational resources and funding.

P. Blumenfeld and colleagues (2000) and B. Fishman and colleagues (2004) describe numerous challenges involved in scaling up computer technology innovations for science education in a large urban school district. The research team created an innovation to promote an inquiry-focused, technology-rich middle school science curriculum and teacher professional development, utilizing a number of design experiments. The researchers developed a conceptual framework and identified issues and concerns in scaling up this innovation in a systemic urban school reform setting.

Blumenfeld and team (2000) developed a diagnostic framework to identify challenges to adopting and sustaining educational innovations in a systemic reform setting. Their framework contains three dimensions. The "policy and management" dimension describes the extent to which established district policies and the management system that carries out those policies are favorable to the demands of the innovation. The "capability" dimension describes the extent to which users have the conceptual and practical knowledge necessary to carry out the innovation. The "culture" dimension describes the extent to which the innovation adheres to or diverges from the existing norms, beliefs, values, and expectations for practice at different levels of the system or organization. The researchers illustrate how the framework exemplifies their experiences in scaling-up. They emphasize the importance of collaborating with teachers and administrators to adapt the innovation so that it is feasible within the constraints of the context, but also true to the underlying premises of the educational approach and the district's own reform agenda. Simultaneous coordination of the different elements of the innovation (e.g., instruction, computer technology, and professional development) is imperative. Furthermore, common understandings and coordination of administrative and organizational rearrangements are also crucial.

On the basis of their experience of scaling-up efforts, Fishman and team (2004) examine why computer technology innovations known to foster deep thinking and understanding have not led to widespread changes in teaching practices or improvement in student learning in K–12 schools. According to the researchers, a key reason is that the research conducted on these technologies has focused on cognitively oriented technology innovations, but it has not yet addressed issues of scalability, sustainability, and usability of technology innovations. Currently, a bridge between the focused research and development of learning technologies and the broad implementation of these innovations in schools is lacking. While it is essential to continue to explore cutting-edge technologies that may not be ready for widespread use in schools, and to conduct basic research on the potential contributions of technology to student learning and understanding, it is also important to engage in systemic research on multiplying the use of technology innovations in schools.

Lee and Luykx (2005) map out major difficulties that arise in scaling-up efforts, specifically with regard to elementary school science and students' linguistic, cultural, and socioeconomic diversity. As the intervention has been taken to scale over the years, policies and practices promoting high-stakes assessment and accountability in reading, writing, and mathematics have become ever more prevalent. Despite the overall effectiveness of the intervention with regard to student achievement in science, as well as change in teachers' beliefs and practices (described elsewhere in this book), the project continues to experience challenges that hinder its full implementation.

First, scaling-up in contexts of student diversity (recalling that virtually any scaling-up intervention involves a diverse student population) requires considerable conceptual work. To date, the knowledge base needed to translate an educational innovation that articulates science disciplines with student diversity into self-sustaining educational policy and practice remains limited. Second, standards-based instruction and accountability policies in a growing number of states reinforce the mainstream view by which nonmainstream groups are expected to assimilate to the dominant culture. Additionally, more states are shifting toward "English only" policies that do not consider students' actual or potential resources in their home languages. Finally, scaling-up efforts should confront the practical, day-to-day challenges that characterize inner-city schools, including high rates of administrator turnover and teacher attrition, disproportionate numbers of beginning or inexperienced teachers, high rates of student mobility, inadequate infrastructure, and insufficient funding. It may be difficult to ensure the sustainability of an innovation in elementary science education beyond the research period, if insufficient resources or instructional time are allocated for science. The researchers conclude that given the existing tensions and conflicts around educational policies and practices for nonmainstream students in elementary school science, it is important to identify factors affecting the feasibility and fidelity of implementation of scaling-up efforts, and to make these factors a subject of open debate and analysis in the educational research community.

Accountability Policy

Under the current education reform (NCLB Act) with its emphasis on standards-based instruction, state content standards or curriculum frameworks offer guidelines for school curricula and classroom instruction (Cohen & Hill, 2000; Knapp, 1997; Smith & O'Day, 1991). After almost a decade of high-stakes assessment in reading, writing, and mathematics, state and national policies are now shifting to include science and social studies as well. This trend coincides with the planned federal policy on science accountability within the NCLB Act, scheduled to take effect in 2007. Policymakers promote high-stakes assessment as one way of addressing what has historically been a dual challenge for schools: high achievement in the academic subject areas and educational equity for an increasingly diverse student population (Darling-Hammond, 1996; McLaughlin et al.,1995). The inclusion of all students in such tests is an attempt to ensure that they all have at least nominal opportunities to encounter the same content. However, critics charge that high-stakes assessments in effect stratify students by race/ethnicity, SES, and linguistic background, given that these factors largely determine students' access to learning opportunities.

As states increasingly turn to standardized tests for accountability, high-stakes assessment influences instructional practices both in subject areas being tested and in those that are not tested. When science is not part of accountability measures, it is taught minimally in the elementary grades (Knapp & Plecki, 2001; Spillane et al., 2001). Schools serving low-income and ELL students are pressed to ensure students' basic proficiency in standard English literacy and numeracy, often at the expense of other subjects such as science.

When science is part of accountability measures, teachers are pressed to model their teaching practices after the demands of high-stakes assessment. If science assessments emphasize basic skills and low-level knowledge of large amounts of content, scaling-up efforts to promote high academic standards may face resistance (Bianchini & Kelly, 2003). On the other hand, if science assessments are designed to measure deep under-standing and higher-order thinking about key topics or concepts, prac-titioners are likely to be more receptive to innovative science programs, especially if the programs are intentionally aligned with such assess-ments. However, even in such conditions, the pressure to cover increas-ing amounts of content in preparation for high-stakes assessment and accountability may force teachers to choose between depth and breadth of coverage.

The current literature on high-stakes assessment and accountability in science education is limited (Deboer, 2002; Wideen et al., 1997), particu-larly with regard to studies involving nonmainstream students (Secada & Lee, 2003; Settlage & Meadows, 2002). J. Settlage and L. Meadows (2002) examined the influences of standards-based reform and standardized test-ing as experienced by four science teachers in urban schools in Cleveland, Ohio, and Birmingham, Alabama. Three of the four teachers were African American; two taught in elementary schools and two in high schools. All four were exemplary teachers, respected by their peers, dedicated to their students, and determined to improve themselves and the systems in which they worked. By means of classroom observations and teacher interviews, the researchers identified several unintended and harmful consequences of standards-based reform and standardized testing: (a) the erosion of teacher professionalism, (b) the disruption of interpersonal relationships between teachers and their students, (c) the trivialization of science instruction, and (d) the adoption of an educational "triage" mentality. The researchers pro-pose strategies that university-based science educators might employ, such as assisting teachers in resisting these reform-induced perils and treating the conditions faced by urban schools as a substantive aspect of science teacher education.

W. G. Secada and Lee (2003) examined teachers' conceptions and prac-tices around mathematics and science (curriculum, instruction, and assess-ment) in relation to student diversity (ability, gender, culture, language, and SES). They compared highly effective and typical elementary schools in two school districts with high levels of racial/ethnic and linguistic diver-sity. The study involved two 4th-grade teachers from each school within each district, for a total of 16 schools and 32 teachers. Within each district, 5 highly effective and 3 typical schools were selected on the basis of stu-dent achievement on that district's mathematics and science achievement tests. The study utilized classroom observations, teacher interviews, and student work samples.

The study tested the hypothesis that compared to teachers in typi-cal schools in both districts, teachers at highly effective schools would demonstrate more effective conceptions and practices with regard to math/science and more equitable conceptions and practices with regard to student diversity. The results showed very little evidence that the instructional practices at highly effective schools were more effective, or that learning environments were more equitable. However, there was compelling evidence that teachers at highly effective schools had better-developed and better-articulated conceptions of mathematics and science, and also of student diversity. Teachers at highly effective schools tended to state that they valued their students' informal strategies for doing science and mathematics, looked for and found potential to do science and mathematics in their low-achieving students, and/or relied on their racial/ethnic and linguistic similarities with their students to encourage the students to engage in academic tasks.

Differences in high-stakes assessments between mathematics and science and between the two school districts offer insights that are worth further exploration. In mathematics, teachers in both districts had to prepare their students for standardized tests. This seemed to force teachers to focus on a set of skills and knowledge expected on the tests, particularly in the school district with high accountability measures based on statewide assessments. In science, teachers in one district had to prepare students for a standardized test emphasizing science process skills, whereas science would not be tested in the other district until the year after the study was conducted. Teachers in both districts promoted scientific processes and hands-on activities, although for different reasons (in preparation for the standardized science test in one district, and without the pressure of a standardized science test in the other). In both districts, students had opportunities to engage in scientific processes and hands-on activities, compared to drill and practice of basic mathematical skills and a rigid problem-solving process. These results indicate that what is measured in high-stakes assessments strongly influences what is taught in mathematics and science classrooms.

Discussion

The literature on school organization in science education is limited to urban contexts, with no studies focusing specifically on ELL students. Some studies used descriptive research methods (e.g., Gamoran et al., 2003; Spillane et al., 2001), whereas others examined the relationships between school restructuring and student achievement using large databases (e.g., Lee & Smith, 1993, 1995; Lee et al., 1997).

Despite the limited literature, major findings seem to emerge with regard to effective and equitable school organization for nonmainstream students. Schools that promote high academic achievement for all students (effective) while also narrowing achievement gaps (equitable) display practices consistent with the restructuring movement, have a strong academic focus, offer a highly academic curriculum with limited use of tracking, create smaller organizational units ("schools within schools"), and support strong professional communities of teachers with a focus on the quality of instruction. These schools also emphasize strategic use of human, social, and material resources and distributed leadership among administrators and teachers.

In response to recent educational policies promoting standards-based instruction, systemic reform, and accountability, there is an emerging literature on educational policies in science education. Like those focusing on school organization, these studies deal exclusively with urban contexts, and none focuses specifically on ELL students. The research has been strongly influenced by recent federal policy initiatives. When there was substantial funding for systemic reform of mathematics and science education (e.g., the NSF-funded Statewide Systemic Initiative and the Urban Systemic Initiative), research (mostly in the form of evaluation) focused on the impact of systemic reform on student outcomes, science curriculum, teacher professional development, and infrastructure for systemic change (e.g., Kim et al., 2001a, 2001b). Under the current policy initiative for research-based practice (i.e., "what works"), there is a new focus on taking effective practices to scale and testing the impact of large-scale initiatives across a range of educational settings (Blumenfeld et al., 2000; Fishman et al., 2004). Additionally, as school science will become part of federal accountability policy starting from 2007, research on the influences of high-stakes assessment and accountability on science instruction (particularly in urban schools with high proportions of nonmainstream students) is expected to emerge (Secada & Lee, 2003; Settlage & Meadows, 2002).

Current educational policies are unique in terms of federal and state mandates on accountability in state education systems, which have historically been driven by local initiatives. Research in science education needs to make conceptual and methodological advances to address policy issues accordingly, especially as regards the scaling-up of educational innovations beyond even the district- or statewide level. To answer the question of what constitutes "best policies and practices" in science education for diverse student populations, rigorous attention to the challenges facing schools and teachers in articulating science disciplines with nonmainstream students' linguistic and cultural experiences is urgently needed (Lee & Luykx, 2005).

9

School and Home/Community Connections

Science achievement gaps among racial/ethnic and social class groups have been extensively documented, and several studies have examined the influences of families and home environments on students' science achievement. A challenge facing many schools is the lack of connection between schools and students' homes and communities. Students are more likely to disengage from schooling if they see it as irrelevant and meaningless to their lives beyond school. Yet students bring to the science classroom "funds of knowledge" from their communities that can serve as resources for academic learning, and academic learning is mediated by highly articulated tasks and activities that occur in the social contexts of day-to-day living, whether or not the school chooses to recognize this (Moll, 1992; Vélez-Ibáñez & Greenberg, 1992). The small number of studies on connections between science learning and students' families and communities can be grouped into three areas: (a) the influence of families and home environments on students' science learning, (b) school science and community connections, and (c) science learning among homeless children.

Families and Home Environments

There is clear evidence that family support influences children's achievement, attitudes, and aspirations, even after student ability and family SES are taken into account (Epstein, 1987; Oakes, 1990). R. A. Schibeci and J. P. Riley (1986) examined the influence of five background variables (i.e., race, gender, home environment [e.g., having an encyclopedia in one's home], amount of homework, and parents' education) on science achievement and attitudes. Using the LISREL method for statistical analysis, the study involved two random samples, comprising 350 and 323 students, from the population of more than 3,000 17-year-old students who participated in the 1976–1977 NAEP survey. The results indicate that home environment, homework supervision, and parent's educational background had substantial influence on students' science achievement and/or attitudes. Additionally, White students scored higher on average in achievement than other racial/ethnic groups. (The article does not provide clear and specific information about results involving science achievement and attitudes for each of the five variables.)

Using the NELS:88 data, S. Peng and S. Hill (1994) examined the influence of home, school, and student factors that differentiated high- and low-achieving minority students (i.e., African American, Hispanic, and Native American and Native Alaskan students) in science and mathematics. Certain factors correlated with science achievement, regardless of racial/ethnic and cultural background. In terms of home influences, high achievers were more likely to come from families with more learning materials and resources. Their parents were more likely to be in highly skilled occupations that provided appropriate role models for science and mathematics learning. Their parents also had higher educational expectations for their children and thus probably created academic pressure and support for them directly and indirectly. These results suggest that it is the economic and educational marginalization associated with racial/ethnic minority status, rather than students' racial/ethnic or cultural background per se, that negatively affects minority students' science achievement.

F. M. Smith and C. O. Hausafus (1998) examined those aspects of family support that had the most influence on low-SES, ethnic minority students' mathematics and science learning. Particularly, they examined the relationship of the mother's support to 8th-grade students' scores on standardized tests in mathematics and science. The student sample included children participating in a special cooperative partnership program between a university and an urban community school district. The 80 students in the sample included 32 recent immigrants from Southeast Asia, 28 African Americans, 17 Hispanics, and 3 Native Americans. Given the low-SES backgrounds of the participants, the research considered ways that parents could be "supportive," even though they might not be able to be "active." Mothers of 80 students responded via telephone with regard to their own behavior, the physical environment of the home, and their attitudes toward science and mathematics. Data were analyzed using multiple regressions and other statistical methods. The results indicated that students had higher test scores if parents helped them see the importance of taking advanced science and mathematics courses, set limits on TV watching, and visited science/mathematics exhibits and fairs with their child. The results are important in the sense that these activities do not require parental knowledge of science and mathematics, areas in which parents often feel inadequate. Instead, parents can be supportive by communicating and enforcing high expectations for achievement with their children.

M. Callanan (2000) offers an in-depth look at the role of parents and teachers in promoting young children's "science talk," including causal questions and explanations for scientific phenomena. The research addressed two sets of questions. Those focusing on home environments included (a) What does science talk look like in everyday conversations in families of Mexican descent? and (b) How do these conversations differ depending on the educational backgrounds of the parents? Those focusing on school environments included (a) What does science talk look like in preschool and early elementary classrooms? and (b) If teachers knew more about the characteristics of science talk in students' families, how might they make use of this information as a "fund of knowledge"? The researcher examined parent–child interactions among 48 families of Mexican descent living in northern California. The research employed qualitative methods, particularly videotaped data from homes and classrooms. The results indicated that parents and children engaged in many conversations about scientific topics, regardless of parents' educational backgrounds. The results also indicate that parents' educational background influenced their explanatory talk with their children. For example, in formal settings (e.g., museums), parents with more formal education were more likely to talk to their children about the content of the exhibits, whereas parents with less formal education were more likely to engage their children in general conversations about what museums were like. In contrast, in settings that were familiar to both groups of families (e.g., the home), parents' educational backgrounds did not seem to make a difference to their conversations with children about science topics. For example, in an at-home flotation activity in which families were given a large tub of water and a number of different objects, parents from both groups made predictions and supported them with scientifically relevant ideas. Furthermore, when teachers learned about the characteristics of science talk in students' families, virtually all of them were motivated to make links between children's own ideas and science concepts, and some were quite successful in making use of children's ideas in classroom settings.

School Science and Community Connections

Several studies investigated intervention programs to help students recognize the meaning and relevance of science and make connections between school science and their communities. L. Hammond (2001) described collaborative efforts in which mentor and preservice teachers worked together with immigrant students and their families to explore options for elementary science and other subject areas. A team of bilingual/multicultural teacher educators worked in conjunction with teacher researchers, most of whom were immigrants themselves, representing a variety of communities, languages, and cultures. Through these efforts, children, teachers, and student teachers gathered community "funds of knowledge" about the science topics to be studied, and then incorporated this knowledge into instruction by using parents as experts and creating "community books." The community-generated materials paralleled and complemented standards-based curricula, although science topics that had natural significance in particular communities were used as a starting point. Using critical ethnography, the study focused on Central Asian immigrants, including Mien and Hmong families, in an urban school district in California. In describing the process of building a Mien American house, the study illustrates how a new kind of "multiscience" can emerge by drawing on participants' funds of knowledge, and how such multiscience can be made accessible to all collaborating members and responsive to school standards.

L. M. Bouillion and L. M. Gomez (2001) explored a form of "connected science" as a way to provide all students with opportunities for meaningful and intellectually challenging science learning. In connected science, real-world problems (i.e., current, unresolved, and consequential) and school–community partnerships were used as contextual scaffolds for bridging students' community-based knowledge and school-based knowledge. Their case study examined the potential of these scaffolds for connected science with a team of elementary teachers at one elementary school. The study used the student-identified problem of pollution along a river near the students' predominantly Mexican American neighborhood. The community partners included parents, scientists, and local community organizations. The study focused on how diverse forms of science knowledge and experience could be brought together in support of student learning, particularly in urban settings in which students' home culture often differs from or may even conflict with the culture of power as represented in school science. The results indicated that these diverse forms of knowledge supported project activities and enhanced students' science learning, interest, and efficacy.

J. Rahm (2002) described an inner-city youth gardening program and the kinds of science learning opportunities it supported. Using a qualitative case study approach, the research involved middle school inner-city students who were at risk of dropping out of school and had few opportunities to engage in other extracurricular activities or summer programs. The research took place in a 4-H community youth program that ran through the summer, in which the students earned money by participating in gardening and selling the produce in a community market. Particularly, the study examined the ways in which the garden environment and the experiential nature of the program supported youth-initiated actions and talk, while also enabling connections among science, community, and work. Additionally, the study emphasized the value of a science that emerged from participants' engagement in activities they deemed valuable, meaningful, and authentic. Some students volunteered to continue their gardening projects after the summer program was over. The results indicated that the gardening program, though it did not have science as its primary goal, provided valuable science and mathematics learning opportunities that were meaningful, relevant, and real to the students. The results also highlighted the educational value of a science practice that was driven by students, rather than being imposed on them, and that provided opportunities for the integration of science, community, and work.

Science Learning of Homeless Children

The research program led by A. Calabrese Barton has examined issues of science teaching and learning with urban homeless children, who are most at risk for receiving an inequitable education (or no formal education at all). The research program employs critical ethnography, conceived as a methodology that emerges collaboratively from the lives of the researcher and the researched and is based on a political commitment to the struggle for liberation and the defense of human rights (Calabrese Barton, 2001).

Grounded in critical, postmodern feminism, the research program argues that if science for all students is to be a reality, the reflexive nature of the relationship between science and student diversity should be explicitly articulated (Calabrese Barton, 1998a). This view of "science for all" requires attention to the social, historical, political, and physical contexts of children's lives, epistemologically decentering science and positioning it in a dynamic relation of multiple realities. The view shifts from the traditional paradigm where science lies at the center, as a target to be reached by students at the margins, to one of inclusion where students' experiences and identities remain in tension with the study of the world.

Calabrese Barton (1998b) worked with urban homeless youth in an after-school science program conducted in a homeless shelter. The students took the lead in planning activities, documenting their explorations, and constructing meaning from their findings. The role of the researcher, as teacher, was to validate the students' experiences by using these experiences as the starting point for their explorations, and to help them locate questions in their experiences and find ways to critically explore those questions. Through the sharing of their personal theories of the local community's pollution or the shelter's policies around food, the students in the program used their own lived experiences to define science in terms of both practice and content. Throughout the teaching and learning process, students' identities remained a central focus of a pedagogy conceived on democratic principles. The researcher argues that urban homeless children carry to school with them a set of struggles not reflected in the typical science curriculum, and that the pedagogical questions of identity and representation should be central to their science learning.

Calabrese Barton (1998c) described how homeless children's construction of science was manifested through invention or inventive acts, such as creating a recipe for soup or making a purse out of beads and other supplies. The researcher interpreted the children's inventive acts in terms of three themes: invention as a social act, invention as a recursive and socially linked process, and invention as embodied agency. From a postmodern perspective, the researcher argues that the connection between scientific/technological knowledge and power should be carefully examined. She further argues that science/technology education should actively and continually deconstruct the "master narrative" of modern science, encourage individuals and groups to engage in inventive acts as a way to challenge existing social conditions, and enable them to use their lived experiences to challenge imposed definitions of science/technology in and out of schools.

D. Fusco (2001) raised the issue of why youth find informal science experiences (i.e., nonschool-, noncurriculum-based interactions with science in environments such as science centers, museums, zoos, parks, and nature centers) fun and relevant to their futures. The researcher worked with teenagers from homeless families in an after-school project that involved urban planning and community gardening. She concluded that in this community-based science project, science became relevant or "real" to the students because (a) it was created from their own concerns, interests, and experiences inside and outside of science; (b) it was an ongoing process of researching and then enacting ideas; and (c) it was situated within the broader community.

Fusco and Calabrese Barton (2001) further argued that this community-based science project could serve as the context for performance assessment, as students collectively created the community garden and produced a written document about their project. Students' understandings of science content and the nature of science were supported by authentic and meaningful practice. The teens were active producers of a science that made sense to them, served a communal purpose, and drew upon their interests and strengths. Conceptions of science and science content emerged from real-world connections. Here, students' science learning was supported by a vision of science as socially oriented rather than task oriented. The enactment of science was situated holistically and historically, and the written document was representative of the totality of students' achievements.

Discussion

There is a small but emerging body of literature on the connections between school science and students' home and community environments. Except for the three studies that employed correlational methods to examine the relationship between family/home environments and science achievement (Peng & Hill, 1994; Schibeci & Riley, 1986; Smith & Hausafus, 1998), all the other studies used qualitative research methods. The work by Hammond (2001) and Calabrese Barton and colleagues with homeless children were characterized by a critical ethnographic perspective focusing on the articulation of science with social and political power.

This emerging literature has yielded some key findings. Traditionally, this area of research looked at how the family and home environments of nonmainstream students measured up to the expectations and practices of the mainstream. The results were interpreted in terms of "deficiencies" in students' family and home environments, as compared to their mainstream counterparts. Recent research, in contrast, has identified resources and strengths in the family and home environments of nonmainstream students. For example, differences between the science achievement of mainstream and nonmainstream groups are attributed largely to disparities in prior knowledge about science topics and familiarity with task settings, rather than differences in general ability or the cultural "richness" of students' home environments (Callanan, 2000). Even parents with limited education or limited science knowledge can promote their children's science learning by communicating and enforcing high expectations for science achievement (Smith & Hausafus, 1998).

Several studies provided detailed descriptions of students' engagement and learning as they participated in intervention programs connecting school science with their community environments (Bouillion & Gomez, 2001; Hammond, 2001; Rahm, 2002; the work by Calabrese Barton and colleagues). The results consistently indicate that students gained a better understanding of science, recognized the relevance of science to their personal lives, and developed interest and agency in science as either a direct or indirect outcome of the interventions. These studies argue that informal science experiences are critical for learning science and that school science should be reconceptualized in such a way that students' lived experiences and identities are given a central role.

CONCLUSIONS AND A RESEARCH AGENDA

This synthesis follows seminal works on diversity and equity in science education, including those by J. Oakes (1990), M. M. Atwater (1994), and S. Lynch (2000). It differs from these previous works in several important respects. First, it examines the relationship between students' science outcomes, broadly defined, and mechanisms or factors thought to affect outcomes. Particularly, it attempts to provide insights into why gaps in science outcomes among racial/ethnic, cultural, linguistic, and socioeconomic groups have persisted over the years, in the hope that such insights may point the way toward eliminating those gaps. Second, it is based predominantly on peer-reviewed journal articles and guided by other criteria whose aim is to ensure the methodological rigor of the research considered herein. Third, considering that a majority of the studies on diversity and equity in science education have been conducted and published since the mid-1990s, it offers the most current review of the literature. Finally, a major goal of this synthesis is to offer recommendations for a future research agenda.

10

Conclusions

We offer conclusions regarding two areas: (a) key features of the literature with regard to theoretical perspectives and methodological orientations, and (b) key findings about factors related to science outcomes of non-mainstream students. Future research should address current limitations in theory building and in research methods. Future research should also pursue those areas that demonstrate promising findings with regard to improving science outcomes and narrowing gaps among diverse student groups, as well as those areas or topics that have been largely ignored in the current literature.

Key Features of the Literature

Research on diversity and equity in science education is a new and developing area. Studies have been conducted from a wide range of theoretical and disciplinary perspectives, ranging from cognitive science, to cross-cultural, to sociopolitical. They have utilized a variety of research methods, ranging from experimental designs, to surveys, to critical ethnography. Experimental studies were rare, relative to the many studies using qualitative methods. No metanalysis of statistical research studies was found in the literature.

Given that student diversity and science education is an emerging area of research, there are many conceptual reviews or articles explicating particular issues and framing such issues for research pursuits. There are only a small number of programmatic lines of research carried out by research teams, and these research programs have emerged only in recent years. The majority of studies are small-scale, descriptive research conducted as single studies by individual researchers. There are only a small number of intervention-based studies, and relatively few of these are on a large scale.

There are substantial bodies of literature on some topics (e.g., multicultural science, worldviews as they relate to the epistemology of science or science learning, science instruction), but only limited research on others (e.g., science assessment, influences of school organization and educational policy on science education, home and community connections to school science). Elementary and secondary school education are both well represented in the literature. Specific science disciplines or topics usually serve as the research context, rather than the focus, and research findings concerning particular science disciplines or topics are assumed to be applicable to other disciplines or topics.

The relationship between students' science outcomes (particularly achievement data) and educational processes or mechanisms is tenuous in most studies. Studies focusing on educational processes or mechanisms often do not report student outcome data, and studies linking student outcomes to causal mechanisms or factors are even fewer. Studies reporting on the impact of intervention programs on achievement gaps among racial/ethnic, cultural, linguistic, and socioeconomic groups are scarce. Since outcome measures are usually developed independently by researchers for use in a specific study, it is difficult to compare outcomes across studies.

This synthesis is an attempt to consider the relationships of race/ethnicity, culture, language, and SES to science education in a broad sense. With some exceptions, studies generally treat these variables separately and fail to examine the intersections among them (see the discussion about research methodology in Luykx & Lee, in press). For example, studies focusing on poverty in urban education often do not consider students' language or culture in a systematic manner, and studies focusing on language and/or culture in the classroom seldom consider the broader social context in which classroom interactions are situated. As J. L. Lemke (2001) noted:

There seems to be some tendency in the literature to apply only one type of sociocultural analysis for each social group, neglecting the role of the others. For example, in the U.S. literature, we hear far more about race in relation to African-Americans than we do about language or social class; far more about language in the case of Hispanic groups than we hear about race or class; and far more about culture for Asian-Americans or Native Americans than about race, language, or class. (p. 303)

On the other hand, studies often fail to distinguish (conceptually and/or methodologically) the variables that tend to co-occur, e.g., race/ethnicity with culture, SES with prior science knowledge, and literacy with English proficiency. Rather than conflating such variables or treating them in isolation, future research should aim to examine their intersections more systematically.

Reflecting the incipient nature of most research on science education and student diversity, "science education researchers are not often enough formally trained in the disciplines from which sociocultural perspectives and research methods derive" (Lemke, 2001, p. 303). The degree of theoretical and methodological sophistication with which cultural and linguistic issues are treated in the science education literature is particularly uneven. While some researchers have gone beyond the limits of narrowly defined educational research, incorporating important concepts and research findings from related fields like educational sociolinguistics and cultural anthropology, others neglect even the most basic principles of these fields. The disciplinary "tunnel vision" of much science education research has frequently given rise to research designs that are fundamentally flawed, and interpretations that are markedly ethnocentric or uninformed with regard to cultural and linguistic processes. In some studies, the methods used for sampling or assessment compromised the rigor of the data collection; in others, the information provided on language treatment in the context of science instruction was insufficient to support any specific conclusion on the effects of linguistic factors on science outcomes. However, a very few studies (conducted outside the United States) display greater methodological rigor and theoretical depth in this regard. They demonstrate a commendable attention to the sociolinguistic context of science education, including features of language policy and "language ecology" that exert a powerful influence on instructional processes (Cleghorn, 1992; Kearsey & Turner, 1999).

Considering that ELL students are faced with the challenge of acquiring oral and written English along with science knowledge, and that this challenge manifests itself in all the areas touched upon in this book, the synthesis has treated studies involving ELL students separately within each section. Studies reporting student outcomes in both science and literacy are rare. Only two studies, by Amaral and colleagues (2002) and Lee and colleagues (2005), compared student achievement in science and literacy at different levels of English proficiency. Even in these rare cases, fundamental questions about the role of English and students' home language in science instruction, learning, and assessment tend to be skimmed over or ignored, thus hindering the interpretation of research findings.

Another notable aspect of this research is how little informed it is by research on bilingualism more generally. Few studies undertaken in the United States posit any major instructional role for students' home languages, and the question of whether students are literate in the home language is not addressed. Often, this is because the relevant information is not available to researchers, inasmuch as U.S. schools seldom assess students' literacy in any language other than English. Much of the literature seems to assume that "the language of science" is synonymous with English, and some researchers (Curtis & Millar, 1988; Duran et al., 1998; P. P. Lynch, 1996a, 1996b) seem to assume that English proficiency is a prerequisite to meaningful science learning. Studies conducted outside the United States (e.g., Cleghorn, 1992; Tobin & McRobbie, 1996) are more likely to examine the role that students' home languages play in their science learning. This confirms the powerful influence of national and state language policies on research agendas and programs; in countries with a tradition of mother-tongue schooling (e.g., India), researchers are apparently less constrained in terms of exploring the intersections between school science and linguistic diversity.

A notable trend in science education research is the attempt to implement and test educational interventions on a large scale at the district and state levels. This trend reflects the emphasis on systemic reform in mathematics and science education during the 1990s (primarily through the NSF support) and, more recently, on scaling-up of educational innovations (primarily through the NCLB Act). Not only do these studies face the cultural and linguistic complexities described above; they are also confronted with the day-to-day challenges inherent in large urban education systems. Inevitably, some of the obstacles will not be overcome, even in the most creative research endeavors. Furthermore, the dominant policy context of high-stakes assessment and accountability may limit implementation of ambitious interventions (e.g., long-term and intensive teacher professional development), if the high-stakes assessment emphasizes low-level knowledge or skills or if insufficient resources or instructional time are allocated for science instruction. Such policies may also limit implementation of certain educational programs (e.g., bilingual education programs or performance assessments) due to ideological conflicts or resource constraints. The theoretical underpinnings and methodological rigor in these studies need to be assessed against the policy contexts in which the research is conducted.

Key Findings in the Literature

While science learning is demanding for all students, achievement gaps indicate that it is more demanding for nonmainstream students. Students from all racial/ethnic, cultural, linguistic, and socioeconomic backgrounds come to school with already-constructed knowledge, including their home language and cultural values, acquired in their home and community environments. Such knowledge serves as the framework for constructing new understandings. However, some aspects of students' experience may be discontinuous with science disciplines as traditionally defined in Western modern science. Furthermore, even those experiences of nonmainstream students that could potentially serve as intellectual resources for new learning in science classrooms are generally marginalized from school science.

The education system often fails to provide equitable science learning opportunities for nonmainstream students. Curriculum materials seldom incorporate cultural experiences, analogies, or artifacts representative of nonmainstream groups. Teachers are generally unaware of cultural and linguistic influences on student learning; many do not consider "teaching for diversity" as their responsibility, purposefully overlook cultural and linguistic differences and accept inequities as a given condition, or actively resist multicultural views of learning. Assessment practices are biased in various ways, inasmuch as ELL students are seldom assessed in their home language, low-income students confront test items that are unrelated to their daily lives, and knowledge of the cultural conventions shaping academic discourse is assumed more often than it is taught. All of these assessment practices may result in a major underestimation of nonmainstream students' science knowledge, since they conflate science knowledge with other types of cultural and linguistic knowledge. At the same time, the mediation of science instruction by the cultural and linguistic knowledge of the mainstream serves to reduce science learning opportunities for nonmainstream students. Tracking or "ability grouping" also results in inequitable learning opportunities for different student groups, as nonmainstream students are often placed in lower tracks where content is less challenging, science course offerings are minimal, and expectations of student achievement are lower. The consequences of such policies are especially critical in urban school districts because of the array and scope of the obstacles facing urban schools, and the institutional precariousness under which they operate.

When nonmainstream students are provided with equitable learning opportunities in school or in their communities, they demonstrate academic achievement, interest, and agency. Learning environments that articulate the relation of science disciplines with students' cultural and linguistic practices enable students to capitalize on their experiences as intellectual resources for science learning, and to explore and construct meanings in ways that relate science to their social, cultural, and linguistic identities. Educational policies and practices must also provide support for science learning in the form of both conceptual underpinnings and educational resources. Ideally, students could become bicultural and bilingual border crossers between their own cultural and speech communities and the science learning community, able to perform competently in a variety of contexts.

Although effective learning environments share the principle of articulating students' cultural and linguistic experiences with science disciplines, specific approaches to achieving this goal differ from one theoretical perspective to the next. For example, from a cross-cultural perspective, when students are not from the "culture of power" of the dominant society (e.g., Western modern science), teachers need to make that culture's rules and norms explicit and visible for students and help them make smooth transitions between different cultural contexts (Lee, 2003; Lee & Fradd, 1998).

Additionally, for students who come from backgrounds in which questioning and inquiry are not encouraged or for students with limited science experience, teachers can move progressively along the teacher-explicit to student-exploratory continuum, while students learn to take the initiative and assume responsibility on their own (Fradd & Lee, 1999; Lee, 2002). In contrast, from a cognitive science perspective, there is significant overlap between students' explorations of the natural world and the way science is practiced by scientists (e.g., Rosebery et al., 1992; Warren et al., 2001). Teachers need to understand the complex dynamics between scientific practices and students' everyday knowledge, and facilitate and guide students' investigations of their own questions as they learn to speak, read, write, think, and act as members of a science learning community. A sociopolitical perspective shifts from the traditional paradigm that locates science at the center, as a target to be reached by students at the margins, to a decentered epistemology whereby students' experiences and identities remain in tension with scientific (but just as culturally specific) ways of studying the world. Within this view, the teacher's role is to validate students' experiences by taking them as the starting point for explorations of the natural and social world, and to help students arrive at more rigorous and powerful ways of critically exploring questions relevant to their lives.

The status of science education for diverse student groups needs to be understood within the current policy context of high-stakes assessment, where science is generally not salient. Testing in science is not required by federal policies until 2007, according to the NCLB Act, and has not been part of the accountability system in many states (Council of State Science Supervisors, n.d.). This policy context largely determines learning opportunities in science, particularly for nonmainstream students. Reform-based instructional materials in science education are limited, and science instruction is often neglected, especially for ELL students and students who are perceived to be weak in literacy and numeracy. School funding and resources for science instruction are often overlooked because of the pressure to support core subjects of reading, writing, and mathematics.

Efforts to develop curriculum materials for culturally and linguistically diverse student groups present particular challenges. There is a deep concern over the fact that science curriculum materials tend to exclude the cultural and linguistic experiences of nonmainstream students. Despite such concern, curriculum development efforts for nonmainstream student populations are few and far between. Even when culturally relevant materials are developed and prove effective, their effectiveness may be limited to the particular cultural or linguistic group for which they are designed. Conversely, while materials developed for wide use (particularly computer-based materials) may be implemented across a range of educational settings, local adaptations are essential for such materials to be used effectively, which in turn requires expertise on the part of teachers.

This issue has implications not only for curriculum but also for assessment. Except for NAEP or TIMSS public release items, there are few widely used achievement tests in science. For this reason, research programs often develop their own assessment instruments, and many employ authentic or performance assessments. The resulting variability in assessment practices hinders the comparability of research findings. Additionally, assessments designed for specific cultural and linguistic groups may not be valid or equitable for other groups. Efforts to develop either culturally neutral or culturally relevant assessments each present their own set of difficulties. Either way, ensuring valid and equitable assessments is complicated due to the inevitable influence of external factors (e.g., students' linguistic and cultural experiences) that confound the measurement of the construct being assessed (e.g., scientific knowledge or inquiry).

Science education research occurs and (for the foreseeable future) will continue to occur within the confines of the policies promoting standards-based reform and accountability. The desire for standardization and common outcome measures, for both accountability and research purposes, limits the specificity of educational interventions and research efforts with diverse student groups. In other words, *the goal of maximizing both the generalizability of research and overall student outcomes conflicts with the goal of optimizing research designs and individual student outcomes through contextualized modifications of educational interventions.* Sensitivity to student diversity requires that accommodations be made to meet the needs of particular student groups or individuals, whereas scaling-up involves seeking standardized solutions that are applicable to the greatest number of students. These tensions become more acute in inner-city classrooms where student diversity is greater and educational resources and opportunities are more limited.

11

Research Agenda

Considering that research on diversity and equity in science education is a new and emerging literature, future research can pursue a multitude of issues in a multitude of ways. Virtually all of the areas discussed in this synthesis require further investigation. However, priorities for future research need to be identified in order to produce research outcomes that are rigorous, cumulative, and usable for educational practice. Some of the directions proposed here grow out of that literature which has shown particular promise for establishing a robust knowledge base, whereas others are proposed because the urgent need for a knowledge base in these areas has yet to be fulfilled by the limited research that exists.

Science Outcomes

One area ripe for investigation involves conceptions and measurement of science outcomes. Some research programs, especially those adopting a critical theory perspective, emphasize students' agency with regard to science, rather than more commonly recognized outcome measures based on academic achievement. Conceptions of science outcomes vary widely from one research program to another, and also tend to differ from classroom assessment practices, which continue to emphasize memorization of facts. While science educators (researchers, teachers, policymakers, and others) share the dual goals of improving science outcomes and eliminating gaps, existing research programs often do not address student outcomes, especially quantitative achievement data. Without arguing that such data should be the sole currency of educational research, they can provide an additional perspective that confirms or complicates narrative descriptions about other types of student outcomes, which are common in many research studies.

Lack of emphasis on science in current educational policies means that there are few assessment instruments that are widely used in K–12 science.

This obliges researchers to develop their own assessment instruments, often around authentic or performance assessments that are aligned with the goals of the research. Such instruments may be well tailored to the goals of a specific research project, but limit comparability across studies. In general, the limited range of standardized tests in science makes it difficult to develop a cumulative knowledge base about student achievement in specific science disciplines or topics.

Several issues concerning science achievement deserve special attention. First, more research is needed to examine the effectiveness of educational innovations on achievement gaps among racial/ethnic, cultural, linguistic, and socioeconomic groups. Second, future research needs to consider disaggregation of achievement results for the intersections of race/ethnicity, gender, language, and SES, as well as subgroups within broader racial/ethnic categories. Third, longitudinal analysis of student achievement across several grade levels or beyond the K–12 years is conspicuously absent from the current literature. Finally, future research should explicitly attempt to establish the link between students' learning processes and outcomes.

Student Diversity

Although the studies in this synthesis were selected because of their focus on diversity and equity, many do not address these issues in sufficient depth or complexity. For example, studies focusing on students' race/ethnicity seldom consider socioeconomic strata within racial/ethnic groups, and studies on ELL students' science learning seldom consider the organic link between home language and cultural identity. Future research needs to conceptualize the interrelated effects of race/ethnicity, culture, language, and SES on students' science learning in more nuanced ways. While the intersections among the multiple strands that make up students' (and teachers') identities are being theorized in increasingly sophisticated ways, as are the social forces, processes, and practices that shape students' educational experiences (e.g., Levinson, Foley, & Holland, 1996), these new perspectives have rarely been applied to the area of school science. Given the salience of recent debates concerning the role of religion in school science, future research may also consider religion as yet another dimension of student diversity, with possible effects on students' science learning opportunities.

There is a need for studies that combine cognitive, cultural, sociolinguistic, and sociopolitical perspectives on science learning, rather than focusing on one aspect to the exclusion of others. This will require multidisciplinary efforts bringing together research traditions that have too often been developed in isolation from (or even in opposition to) one another.

Future research on ELL students needs to consider science learning/ achievement, literacy development, and English proficiency as conceptually distinct but interrelated variables, and to operationalize the complex interplay of multiple variables in methodologically rigorous research designs. Science educators and researchers also need to engage more deeply the broad scholarship on classroom discourse, second language acquisition, and literacy development. Though this literature has seldom addressed school science directly, its potential contribution to science education is considerable.

Diversity of Student Experiences in Relation to Science Curriculum and Instruction

A major area of future research should be the cultural and linguistic experiences that students from diverse backgrounds bring to the science classroom, and the articulation of these experiences with science disciplines (Lee & Fradd, 1998; Warren et al., 2001). Researchers should aim to identify those cultural and linguistic experiences of students that can serve as intellectual resources for science learning, as well as those beliefs and practices that may be discontinuous with the specific demands of science disciplines. This will require a balanced view of nonmainstream students' intellectual resources as well as of the challenges they face in learning science.

One notable feature of the research is lack of attention to the prior science instruction that immigrant students may have received in their countries of origin. While ideologies of U.S. superiority in scientific and technical fields may lead teachers or researchers to consider students' out-of-country science experiences as negligible or irrelevant, the mediocre rankings of the United States in international comparisons of school science achievement challenge this assumption. Future research may examine differences in science learning among different immigrant groups, perhaps incorporating relevant theoretical distinctions, such as voluntary versus involuntary immigrants (Ogbu & Simons, 1998), economic versus political refugees, and so on.

An expanded knowledge base around students' science-related experiences could offer a stronger foundation for science curriculum, instruction, and assessment. Students of all backgrounds should be provided with academically challenging learning opportunities that allow them to explore scientific phenomena and construct scientific meanings based on their own cultural and linguistic experiences. At the same time, some students may need more explicit guidance in articulating their cultural and linguistic experiences with scientific knowledge and practices. Teachers (and curriculum designers) need to be aware of students' differing needs when deciding how much explicit instruction to provide and to what degree students can assume responsibility for their own learning (Fradd & Lee, 1999;

Lee, 2002). The proper balance of teacher-centered and student-centered activities may depend on the degrees and types of continuity or discontinuity between science disciplines and students' backgrounds, the extent of students' experience with science disciplines, and the level of cognitive difficulty of science tasks. Further research could examine what is involved in explicit instruction, when and how to provide it, and how to determine appropriate scaffolding for specific tasks and students.

Another area for future research concerns the demands involved in learning science through inquiry (see Bredderman, 1983, for an earlier literature review). Although current reforms in science education emphasize inquiry as the core of science teaching and learning (NRC, 1996, 2000), inquiry presents challenges to all students (and many teachers), as it requires a critical stance, scientific skepticism, a tolerance for uncertainty and ambiguity, and patience. These challenges are greater for students from homes and communities that do not encourage inquiry practices (as reported in the literature on worldviews, described earlier; for detailed discussion, see Fradd & Lee, 1999; Lee, 2002, 2003), for those who have limited experience with school science (Duran et al., 1998; Moje et al., 2001; Songer et al., 2003), and for those who have been historically disfranchised by the social institutions of science and do not see the relevance of science to their daily lives or to their future (Gilbert & Yerrick, 2001; Seiler, 2001; Tobin, 2000). Recent research emphasizes the importance of role models, trust, and personal connections between teachers and students as the starting point for nonmainstream students' participation in science inquiry (Sconiers & Rosiek, 2000). Future research may identify essential aspects of inquiry-based teaching and learning, and how these articulate with the experiences of diverse student groups.

Still another area of research, which currently dominates the landscape of science education in general but has largely been ignored with nonmainstream students, involves the use of computer technology in science curriculum and instruction. A very small number of studies on the use of computer-based programs showed positive science outcomes with ELL students (Buxton, 1999; Dixon, 1995). Additionally, an emerging body of research on science instruction that employs interactive web-based technology shows promising outcomes in inner-city schools (e.g., "LeTUS" by Krajcik, Fishman, Blumenfeld, and their colleagues and the "Kids as Global Scientists" weather program by Songer). Further research in this vein may provide detailed descriptions of how various types of student diversity intersect with the introduction of computer technology into science classrooms, as well as an examination of the impact of large-scale implementation of computer technology on science outcomes across a range of educational settings.

Recent years have seen a growing academic and popular literature on educational settings in which racial/ethnic minority students experience academic *excellence* (Haycock, 1999; Hilliard, 2003; Ladson-Billings, 1994; Sizemore, Brosard, & Harrigan, 1994), in contrast to the usual focus on their academic failure. Literature on ELL students also demonstrates the positive impact of bilingual programs on academic achievement (Thomas & Collier, 1995, 2002). To date, these bodies of literature have not addressed science education specifically, but they constitute a compelling challenge to many of the traditional explanations of "achievement gaps." It is hoped that, in the future, some of these researchers in multicultural and bilingual/ESOL education will examine science achievement specifically. At the same time, given the extraordinary achievements of some innovative schools committed to high-quality instruction for traditionally disenfranchised student groups, science education researchers would be well advised to investigate the processes and outcomes of science instruction in these settings, and the ways in which their successes might be reproduced on a larger scale.

Teacher Education

The literature is replete with accounts of the difficulties that science teachers (who are mostly from mainstream backgrounds) experience in teaching students from nonmainstream backgrounds (see the discussion about teacher education in Chapter 7; also see Bryan & Atwater, 2002). While some teachers have low expectations of nonmainstream students and blame students, their families, or their cultural environments for academic failure, even those teachers who are committed to promoting equity still face challenges related to student diversity in their teaching. These problems are likely to be exacerbated as diversity within the teaching population fails to keep pace with increasing diversity among students (Jorgenson, 2000).

Teachers need not always share the same racial/ethnic background as their students in order to teach effectively. However, effective teachers should have an understanding of students' language and culture and the ability to articulate their students' experiences with science in ways that are meaningful and relevant as well as scientifically accurate. While some teachers may lack the cultural knowledge necessary to identify students' learning resources, even teachers with the relevant cultural knowledge may not recognize it as such or may be unsure of how to relate their students' experiences to science (Lee, 2004). Teacher education programs need to incorporate more in-depth treatment of issues related to student diversity, as well as provide teachers with a more solid foundation in science.

Future research may address how to design teacher education programs to enable preservice and practicing teachers to articulate science disciplines with students' cultural and linguistic practices, particularly when the discontinuities between the two domains are large. Research may also examine how teachers' knowledge, beliefs, and practices evolve as they reflect on ways to integrate these two domains. Additionally, research may examine the challenges involved in bringing about change with teachers who deride or ignore student diversity, resist multicultural views, or reproduce racism through their teaching practice (Ladson-Billings, 1999; Tate, 1997). Some educators (Delpit, 1995; Hilliard, 2003) have argued that the overemphasis on racial/ethnic minority students' academic failure in teacher education programs feeds stereotypes and lowers teachers' expectations of nonmainstream students. From this perspective, research on preparation of K–12 science teachers might examine the relationship between the content of such programs and teachers' subsequent notions of students' science knowledge and abilities.

Teacher education programs that successfully promote fundamental change in teachers' knowledge, beliefs, and practices concerning nonmainstream students tend to involve small numbers of committed teachers over an extended period of time. Effective teacher professional development requires adequate time, resources, and personal commitment on the part of both teachers and teacher educators. Future research may examine what is involved in taking effective teacher education models to scale. Such research may need to engage the conceptual debate between traditional definitions that equate scaling-up with increasing the number of participating teachers, schools, districts, or states and emerging notions that address the depth of change necessary to make change sustainable (Coburn, 2003). Research may also examine stages of going to scale, starting from the design of an innovation, to conducting "efficacy studies," then to conducting "effectiveness studies," and finally to taking the innovation to scale across districts and even states, without the involvement of the original designers (Raudenbush, 2003). Additionally, future research may examine the relationship between resources required and the impact of an innovation in terms of both the number of teachers, schools, districts, or states affected and the sustainability of the impact. Such research is likely to intersect with policies on teacher education at the state or district level; this intersection also deserves further investigation

High-Stakes Assessment and Accountability Policy

Currently, the predominant educational policy, which is particularly consequential for nonmainstream students, involves high-stakes assessment and accountability (Abedi, 2004; Abedi et al., 2004). After almost a decade of high-stakes assessment in reading, writing, and mathematics, more states are now moving to incorporate science and social studies as well. This trend coincides with the planned federal policy on science assessment (within the NCLB Act), according to which science will be included in accountability measures starting from 2007.

This policy change at the federal and state levels may bring about dramatic changes in many aspects of science education. The culture of high-stakes assessment already dominates the teaching landscape. For example, an emphasis on discrete facts and basic skills in high-stakes science assessment discourages teachers from promoting deeper understanding of key concepts or inquiry practices (Settlage & Meadows, 2002). Also, complex issues concerning assessment abound, such as which students are to be included in accountability systems, what assessment accommodations are appropriate, and how content knowledge may be assessed separately from English proficiency or general literacy (O'Sullivan et al., 2003). A basic concern is that ELL students' science achievement is underestimated when they are not allowed to demonstrate their knowledge and abilities in their home language (Solano-Flores & Trumbull, 2003). On the other hand, if science instruction is predominantly in English, simply assessing ELL students in the home language will not guarantee an accurate picture of their science knowledge and abilities.

Future research may examine the impact of policy changes on various aspects of science education. For example, research may address whether teaching for inquiry and reasoning also prepares students for high-stakes assessment (and vice versa), or what the trade-offs are in attempting to achieve both aims when the two are at odds with each other. From an equity perspective, research may examine whether recent policy changes differentially affect students from different backgrounds. More generally, research may examine the institutional, social, and political factors that so often lead educational policies to work at cross purposes to empirically tested "best practices" in science education.

School Science and Home/Community Connections

Students' early cultural and linguistic experiences occur in their homes and communities. If science education is to build upon students' experiences, it requires a knowledge base about the norms, practices, and expectations existing in students' homes and communities. Unfortunately, research on the connection between school science and students' home/community environments is very limited. One consequence is that school science tends to be presented exclusively from the perspective of Western science, without adequate consideration of how science-related activities are carried out in diverse cultures and speech communities. Generally speaking, the daunting task of bridging the worlds of home and school falls on students, who may be forced to choose one at the cost of the other. Given this dilemma, it is not surprising that nonmainstream students are so often underserved, underrepresented, and disenfranchised in science.

Future research should give high priority to examining the science-related "funds of knowledge" existing in diverse contexts and communities. Given the near-exclusive focus in the literature on school science in urban contexts, investigation of the funds of knowledge of nonmainstream students in rural communities (including families of migrant agricultural workers) could be a particularly fruitful area. Such research might focus on how parents and other community members can serve as valuable resources for school-based science learning, or explore various educational approaches in community-based projects that can help students recognize the meaning and relevance of science for their daily lives and for their future.

Closing

In order to achieve the ideal of educational equity in the midst of extensive racial/ethnic, cultural, linguistic, and socioeconomic diversity, school science must value and respect the experiences that nonmainstream students bring from their home and community environments, articulate their cultural and linguistic knowledge with science disciplines, and offer educational resources and funding to support their learning. Policies and practices at every level of the education system should be in concert to provide equitable learning opportunities for all students. The results of this synthesis indicate that when they are provided with such opportunities, nonmainstream students are capable of demonstrating high science achievement, interest, and agency. Thus, we must conclude that many, if not most, of the difficulties faced by these students reside not in themselves, their families, or their communities but in the education systems serving them.

The literature on the intersection of school science and student diversity is currently insufficient to the task of effectively addressing persistent gaps in science outcomes, but it points in some promising directions. Deeper examination of the complex relationships among factors influencing science outcomes, combined with greater attention to the potential contributions of multiple theoretical perspectives and research methods, should produce powerful additions to the existing knowledge base in this emerging field. Just as nonmainstream students must become bicultural and bilingual border crossers in order to gain access to the discourse of science, teachers must learn to cross cultural boundaries in order to make school science meaningful and relevant for all children. Similarly, researchers must also breach the barriers separating different theoretical and methodological traditions if they are to disentangle the complex connections between student diversity and science education.

Appendix

Method for Research Synthesis

In selecting research studies for inclusion in this synthesis, a systematic review of the relevant literature was conducted according to the following parameters:

1. Direct relevance to the topic, that is, studies addressing the intersection between science education and student diversity in terms of race/ethnicity, culture, language, and social class.
2. Studies published since 1982. The landmark for science education reform was the release of the "Science for All Americans" document (AAAS, 1989). The period between 1982 and 2004 spans the years leading up to the release of this document (1982–1989) and more than a decade afterward (1990–2004).
3. Studies conducted within the United States and abroad, but limited to those published in English and focusing on settings where English is the main medium of science education.
4. Studies focusing on science education at the elementary and secondary levels (K–12). Studies involving postsecondary or adult learners are not included.
5. Empirical studies from different methodological traditions, including (a) experimental and quasi-experimental studies, (b) correlational studies, (c) surveys, (d) descriptive studies, (e) interpretative, ethnographic, qualitative, or case studies, (f) impact studies of large-scale intervention projects, and (g) demographics or large-scale achievement data.
6. Literature reviews and conceptual pieces.

Within these parameters, the synthesis includes journal articles, books and book chapters, and technical reports. The process of gathering studies from the various sources was carried out as follows:

First, a search of the Education Resources Information Center (ERIC) database was conducted using the terms "science education" and "school" combined with the following key words: equity, diversity, minority, culture, language, race, and ethnicity. Additionally, "science education" (without "school") was combined with the following key words: multicultural, bilingual, limited English proficient (LEP), English language learner (ELL), English to Speakers of Other Languages (ESOL), English as Second Language (ESL), at-risk, immigrant/immigration, and urban education.

Second, selected journals were reviewed manually, including those supported by the American Educational Research Association (*American Educational Research Journal, Educational Researcher, Review of Educational Research, and Review of Research in Education*), as well as other well-known journals focusing on science education (*Journal of Research in Science Teaching* and *Science Education*), bilingual/TESOL education (*TESOL Quarterly* and *Bilingual Research Journal*), and schooling among nonmainstream populations (*Anthropology and Education Quarterly*).

Third, from these two types of sources, only peer-reviewed journal articles were included. Among these articles, empirical studies, literature review articles, and conceptual pieces were included. Empirical studies were used to report research results, while literature reviews and conceptual pieces were used to frame key issues. Neither practitioner-oriented articles (e.g., teaching suggestions or descriptions of instructional programs, materials, or lesson plans) nor opinion or advocacy pieces unsupported by empirical evidence were included.

Fourth, in terms of methodological rigor, we sought to include studies that presented convincing links between the research questions and the evidence presented, and demonstrated valid conclusions based on the results (Shavelson & Towne, 2002). However, due to the emergent nature of the field, some of the studies were less successful than others at meeting these criteria. We do not engage in detailed critiques of the methodological rigor of individual studies; instead, we provide a general critique of the literature on each research topic.

Finally, based on recommendations by the members of the Science Education and Student Diversity Task Force, the synthesis includes relevant books, book chapters, and technical reports by several organizations with explicit and well-established peer review processes (e.g., National Center for Education Statistics, National Science Foundation) and by national centers supported by the Office of Educational Research and Improvement (now the Institute of Education Sciences) (e.g., Center for Research on Education, Diversity, and Excellence, and National Center for Improving Student Learning and Achievement in Mathematics and Science).

In addition to the studies described here, we also included key studies (both empirical and conceptual) that focus on *either* science education *or* student diversity, to the extent that they have implications for the other focus. Similarly, we chose to include key studies in the general education literature that serve to locate the studies on science education and/or student diversity within the context of larger educational issues. The numbers of included studies within these categories are rather small; they are used to frame research topics and questions in the opening of each chapter.

Once the review was completed, the selected works addressing student diversity in science education were organized according to the following categories: (a) research topic, (b) research method, (c) year of publication, and (d) data source (i.e., journal article, book or book chapter, technical report). This breakdown served to indicate what research topics were studied often or rarely, what research methods were employed most or least frequently, and what trends concerning research topics and methods had emerged over time. The breakdowns of included works by topic, method, and data source are available from the authors.

Several observations can be made about the reviewed literature, with regard to the field in general (points 1, 2, and 3 that follow), research topics (points 4 and 5), and research methods (points 6, 7, and 8). More details are discussed throughout the book.

1. Research on student diversity in science education is a new and emerging field. Most articles were published since the mid-1990s, perhaps spurred by the emphasis on the dual goals of excellence and equity laid out in *Science for All Americans* (AAAS, 1989), followed by *National Science Education Standards* (NRC, 1996).

2. Coverage of articles on topics related to science and diversity varied greatly among scholarly journals in different areas of education. Prominent science education journals, such as the *Journal of Research in Science Teaching* and *Science Education*, have increased their coverage of these topics since the mid-1990s. Each of these two journals produced a number of special issues on these topics in recent years. In contrast, articles on science education were relatively scarce in the education journals focusing on diversity, equity, bilingual/ESOL, and urban education.

3. There were only a small number of programmatic lines of research carried out by research teams over an extended period of time. Most research consisted of single studies conducted by individual researchers.

4. The numbers of studies or articles were uneven across different research topics. There were large bodies of literature on some topics (e.g., multicultural science, worldviews in relation to epistemology of science or science learning, science instruction), but limited research on others (e.g., science assessment, influences of school organization and educational policy on science education, home/community connections to school science).

5. There were many conceptual reviews or articles explicating key issues about student diversity or equity in science education. In contrast, no metanalysis of statistical research studies was found.
6. Most research studies did not include concrete information about student outcomes (whether achievement data or other outcomes) in their results. Notably absent are quantitative achievement results, achievement gaps, or causal factors related to achievement.
7. The level of theoretical and methodological sophistication about student diversity was uneven across research studies. Most studies failed to consider complexities inherent in such constructs as race/ethnicity, culture, language, and social class, or intersections of these constructs as they relate to science education.
8. Many of the empirical studies were conducted using qualitative methods, whereas experimental or quasi-experimental studies were rare. Most studies were exploratory, small scale, or descriptive. Of the small number of intervention-based studies, few were conducted on a large scale.

References

Abedi, J. (2004). The *No Child Left Behind Act* and English language learners: Assessment and accountability issues. *Educational Researcher*, 33(1), 4–14.

Abedi, J., Hofstetter, C. H., & Lord, C. (2004). Assessment accommodations for English language learners: Implications for policy-based empirical research. *Review of Educational Research*, 74(1), 1–28.

Agar, M. (1996). *Language shock: Understanding the culture of conversation*. New York: William Morrow.

Aikenhead, G. S. (1997). Toward a First Nations cross-cultural science and technology curriculum. *Science Education*, 81(2), 217–238.

Aikenhead, G. S. (2001a). Students' ease in crossing cultural borders into school science. *Science Education*, 85(2), 180–188.

Aikenhead, G. S. (2001b). Integrating Western and aboriginal sciences: Cross-cultural science teaching. *Research in Science Education*, 31(3), 337–355.

Aikenhead, G. S., & Jegede, O. J. (1999). Cross-cultural science education: A cognitive explanation of a cultural phenomenon. *Journal of Research in Science Teaching*, 36(3), 269–287.

Akatugba, A. H., & Wallace, J. (1999). Sociocultural influences on physics students' use of proportional reasoning in a non-Western country. *Journal of Research in Science Teaching*, 36(3), 305–320.

Allen, N. J., & Crawley, F. E. (1998). Voices from the bridge: Worldview conflicts of Kickapoo students of science. *Journal of Research in Science Teaching*, 35(2), 111–132.

Amaral, O. M., Garrison, L., & Klentschy, M. (2002). Helping English learners increase achievement through inquiry-based science instruction. *Bilingual Research Journal*, 26(2), 213–239.

American Anthropological Association. (1998). Statement on "race." www.aaanet.org/stmts/racepp.htm. Accessed February 24, 2004.

American Association for the Advancement of Science (AAAS). (1989). *Science for all Americans*. New York: Oxford University Press.

American Association for the Advancement of Science (AAAS). (1993). *Benchmarks for science literacy*. New York: Oxford University Press.

Arellano, E. L., Barcenal, T., Bilbao, P. P., Castellano, M. A., Nichols, S., & Tippins, D. J. (2001). Case-based pedagogy as a context for collaborative inquiry in the Philippines. *Journal of Research in Science Teaching*, 38(5), 502–528.

Atwater, M. M. (1993). Multicultural science education: Perspectives, definitions, and research agenda. *Science Education*, 77(6), 661–668.

Atwater, M. M. (1994). Research on cultural diversity in the classroom. In D. L. Gabel (ed.), *Handbook of research on science teaching and learning* (pp. 558–576). New York: Macmillan.

Atwater, M. M. (1996). Social constructivism: Infusion into the multicultural science education research agenda. *Journal of Research in Science Teaching*, 33(8), 821–837.

Atwater, M. M. (2000). Equity for Black Americans in precollege science. *Science Education*, 84(2), 154–179.

Atwater, M. M., & Riley, J. P. (1993). Multicultural science education: Perspectives, definitions, and research agenda. *Science Education*, 77(6), 661–668.

Atwater, M. M., Wiggins, J., & Gardner, C. M. (1995). A study of urban middle school students with high and low attitudes toward science. *Journal of Research in Science Teaching*, 32(6), 665–677.

Au, K. H. (1980). Participation structures in a reading lesson with Hawaiian children: Analysis of a culturally appropriate instructional event. *Anthropology and Education Quarterly*, 11(2), 91–115.

Au, K. H. (1998). Social constructivism and the school literacy learning of students of diverse backgrounds. *Journal of Literacy Research*, 30(2), 297–319.

August, D., & Hakuta, K. (eds.). (1997). *Improving schooling for language-minority children: A research agenda*. Washington, DC: National Academy Press.

Baker, D., & Leary, R. (1995). Letting girls speak out about science. *Journal of Research in Science Teaching*, 32(1), 3–27.

Ball, D. L., & Cohen, D. K. (1996). Reform by the book: What is – or might be – the role of curriculum materials in teacher learning and instructional reform? *Educational Researcher*, 25(9), 6–8.

Ballenger, C. (1997). Social identities, moral narratives, scientific argumentation: Science talk in a bilingual classroom. *Language and Education*, 11(1), 1–14.

Ballenger, C., & Rosebery, A. S. (2003). What counts as teacher research? Investigating the scientific and mathematical ideas of children from culturally diverse backgrounds. *Teachers College Record*, 105(2), 297–314.

Banks, J. (1993a). Canon debate, knowledge construction, and multicultural education. *Educational Researcher*, 22(5), 4–14.

Banks, J. (1993b). Multicultural education: Historical development, dimensions and practice. In L. Darling-Hammond (ed.), *Review of Research in Education*. Vol. 19 (pp. 3–49). Washington, DC: American Educational Research Association.

Barba, R. H. (1993). A study of culturally syntonic variables in the bilingual/bicultural science classroom. *Journal of Research in Science Teaching*, 30(9), 1053–1071.

Bianchini, J. A., Johnston, C. C., Oram, S. Y., & Cavazos, L. M. (2003). Learning to teach science in contemporary and equitable ways: The successes and struggles of first-year science teachers. *Science Education*, 87(3), 419–443.

Bianchini, J. A., & Kelly, G. J. (2003). Challenges of standards-based reform: The example of California's science content standards and textbook adoption process. *Science Education*, 87(3), 378–389.

Bianchini, J. A., & Solomon, E. M. (2003). Constructing views of science tied to issues of equity and diversity: A study of beginning science teachers. *Journal of Research in Science Teaching*, 40(1), 53–76.

Blake, M. E., & Sickle, M. V. (2001). Helping linguistically diverse students share what they know. *Journal of Adolescent and Adult Literacy*, 44(5), 468–475.

Bland, B. R., & Glasson, G. E. (2004). Crossing cultural borders into science teaching: Early life experiences, racial and ethnic identities, and beliefs about diversity. *Journal of Research in Science Teaching*, 41(2), 119–141.

Blumenfeld, P., Fishman, B. J., Krajcik, J., & Marx, R. W. (2000). Creating usable innovations in systemic reform: Scaling-up technology-embedded project-based science in urban schools. *Educational Psychologist*, 26(3–4), 369–398.

Boone, W. J., & Kahle, J. B. (1998). Student perceptions of instruction, peer interest, and adult support for middle school science: Differences by race and gender. *Journal of Women and Minorities in Science and Engineering*, 4(4), 333–340.

Bouillion, L. M., & Gomez, L. M. (2001). Connecting school and community with science learning: Real world problems and school-community partnerships as contextual scaffolds. *Journal of Research in Science Teaching*, 38(8), 878–898.

Bourdieu, P. (1984). *Distinction: A social critique of the judgment of taste*. London: Routledge.

Brand, B. R., & Glasson, G. E. (2004). Crossing cultural borders into science teaching: Early life experiences, racial and ethnic identities, and beliefs about diversity. *Journal of Research in Science Teaching*, 41(2), 119–142.

Bredderman, T. (1983). Effects of activity-based elementary science on student outcomes: A quantitative synthesis. *Review of Educational Research*, 53(4), 499–518.

Brenner, M. E. (1998). Adding cognition to the formula for culturally relevant instruction in mathematics. *Anthropology and Education Quarterly*, 29(2), 213–244.

Brickhouse, N. (1994). Bringing in the outsiders: Reshaping the sciences of the future. *Curriculum Studies*, 26(4), 401–416.

Brickhouse, N. W., Lowery, P., & Schultz, K. (2000). What kind of girl does science? The construction of school science identities. *Journal of Research in Science Teaching*, 37(5), 441–458.

Brickhouse, N. W., & Potter, J. T. (2002). Young women's scientific identity formation in an urban context. *Journal of Research in Science Teaching*, 38(8), 965–980.

Brown, A. L. (1992). Design experiments: Theoretical and methodological challenges in creating complex interventions in classroom settings. *Journal of the Learning Sciences*, 2(2), 141–178.

Brown, A. L. (1994). The advancement of learning. *Educational Researcher*, 23(8), 4–12.

Bryan, L. A., & Atwater, M. M. (2002). Teacher beliefs and cultural models: A challenge for science teacher preparation programs. *Science Education*, 86(6), 821–839.

Bullock, L. D. (1997). Efficacy of gender and ethnic equity in science education curriculum for preservice teachers. *Journal of Research in Science Teaching*, 34(10), 1019–1038.

Buxton, C. (1998). Improving science education of English language learners: Capitalizing on educational reform. *Journal of Women and Minorities in Science and Engineering*, 4(4), 341–369.

Buxton, C. (1999). Designing a model-based methodology for science instruction: Lessons from a bilingual classroom. *Bilingual Research Journal*, 23(2–3), 147–177.

Buxton, C. (2005). Creating a culture of academic success in an urban science and math magnet high school. *Science Education*, 89(3), 392–417.

Buxton, C. (in press). Creating contextually authentic science education in a "low performing" urban elementary school context. *Journal of Research in Science Teaching*.

Calabrese Barton, A. (1998a). Reframing "science for all" through the politics of poverty. *Educational Policy*, 12(5), 525–541.

Calabrese Barton, A. (1998b). Teaching science with homeless children: Pedagogy, representation, and identity. *Journal of Research in Science Teaching*, 35(4), 379–394.

Calabrese Barton, A. (1998c). Examining the social and scientific roles of invention in science education. *Research in Science Education*, 28(1), 133–151.

Calabrese Barton, A. (2001). Science education in urban settings: Seeking new ways of praxis through critical ethnography. *Journal of Research in Science Teaching*, 38(8), 899–917.

Callanan, M. (1997, 1998, 1999, 2000). *Performance reports on "At-risk preschoolers' questions and explanations: Science in action at home and in the classroom."* Santa Cruz, CA: Center for Research on Education, Diversity & Excellence.

Campbell, J. R., Hombo, C. M., & Mazzeo, J. (2000). *NAEP 1999 trends in academic progress: Three decades of student performance (NCES 2000–469)*. Washington, DC: U.S. Department of Education, National Center for Education Statistics.

Carter, L. (2004). Thinking differently about cultural diversity: Using postcolonial theory to (re)read science education. *Science Education*, 88(6), 819–836.

Catsambis, S. (1995). Gender, race, ethnicity, and science education in the middle grades. *Journal of Research in Science Teaching*, 32(3), 243–257.

Chin-Chung, T. (2001). Ideas about earthquakes after experiencing a natural disaster in Taiwan: An analysis of students' worldviews. *International Journal of Science Education*, 23(10), 1007–1017.

Chipman, S. F., & Thomas, V. G. (1987). The participation of women and minorities in mathematical, scientific, and technical fields. In E. Z. Rothkopf (ed.), *Review of Research in Education*, Vol. 14 (pp. 387–430). Washington, DC: American Educational Research Association.

Cleghorn, A. (1992). Primary level science in Kenya: Constructing meaning through English and indigenous languages. *Qualitative Studies in Education*, 5(4), 311–323.

Cobern, W. W. (1989). A comparative analysis of NOSS profiles on Nigerian and American preservice, secondary science teachers. *Journal of Research in Science Teaching*, 26(6), 533–541.

Cobern, W. W. (1991). *Worldview theory and science education research* (NARST Monograph, Number 3). Kansas State University, KS: The National Association for Research in Science Teaching.

Cobern, W. W. (1996). Worldview theory and conceptual change in science education. *Science Education*, 80(5), 579–610.

Cobern, W. W., & Loving, C. C. (2001). Defining "science" in a multicultural world: Implications for science education. *Science Education*, 85(1), 50–67.

Coburn, C. E. (2003). Rethinking scale: Moving beyond numbers to deep and lasting change. *Educational Researcher*, 32(6), 3–12.

Cochran-Smith, M. (1995a). Color blindness and basket making are not the answers: Confronting the dilemmas of race, culture, and language diversity in teacher education. *American Educational Research Journal*, 32(3), 493–522.

Cochran-Smith, M. (1995b). Uncertain allies: Understanding the boundaries of race and teaching. *Harvard Educational Review*, 65(4), 541–570.

Cohen, D. K., & Hill, H. C. (2000). Instructional policy and classroom performance: The mathematics reform in California. *Teachers College Record*, 102(2), 294–343.

Contreras, A., & Lee, O. (1990). Differential treatment of students by middle school science teachers: Unintended cultural bias. *Science Education*, 74(4), 433–444.

Costa, V. B. (1995). When science is "another world": Relationships between worlds of family, friends, school, and science. *Science Education*, 79(3), 313–333.

Council of State Science Supervisors. (n.d.). Science education assessment: CSSS state assessment information. http://csss.enc.org/assess.htm. Accessed March 7, 2004.

Cuevas, P., Lee, O., Hart, J., & Deaktor, R. (2005). Improving science inquiry with elementary students of diverse backgrounds. *Journal of Research in Science Teaching*, 42(3), 337–357.

Cunningham, C. M., & Helms, J. V. (1998). Sociology of science as a means to a more authentic, inclusive science education. *Journal of Research in Science Teaching*, 35(5), 483–499.

Curtis, S., & Millar, R. (1988). Language and conceptual understanding in science: A comparison of English and Asian language speaking children. *Research in Science and Technological Education*, 6(1), 61–77.

Damnjanovic, A. (1998). Ohio Statewide Systemic Initiative (SSI) factors associated with urban middle school science achievement: Differences by student sex and race. *Journal of Women and Minorities in Science and Engineering*, 4(2–3), 217–233.

Darling-Hammond, L. (1996). The right to learn and the advancement of teaching: Research, policy, and practice for democratic education. *Educational Researcher*, 25(6), 5–17.

Davis, K. S. (2002). Advocating for equitable science-learning opportunities for girls in an urban city youth club and the roadblocks faced by women science educators. *Journal of Research in Science Teaching*, 39(2), 151–163.

DeAvila, E. A., Duncan, S. E., & Navarrete, C. J. (1987a). Cooperative learning: Integrating language and content-area instruction based on *Finding Out/Descubrimiento* (FO/D). National Clearinghouse for Bilingual Education. http://www.ncela.gwu.edu/ncbepubs/classics/trg/02cooperative.htm#deavila1987. Accessed in September 2005.

DeAvila, E. A., Duncan, S. E., & Navarrete, C. J. (1987b). *Finding out/Descubrimiento. Teacher's resource guide.* Northvale, NJ: Santillana Publishing.

Deboer, G. E. (2002). Student-centered teaching in a standards-based world: Finding a sensible balance. *Science & Education*, 11(4), 405–417.

Delpit, L. (1988). The silenced dialogue: Power and pedagogy in educating other people's children. *Harvard Educational Review*, 58(3), 280–298.

Delpit, L. (1995). *Other people's children: Cultural conflict in the classroom.* New York: W. W. Norton.

Delpit, L. (2003). Educators as "seed people" growing a new future. *Educational Researcher*, 32(7), 14–21.

Desimone, L. M., Porter, A. C., Garet, M. S., Yoon, K. S., & Birman, B. F. (2002). Effects of professional development on teachers' instruction: Results from a three-year longitudinal study. *Educational Evaluation and Policy Analysis*, 24(2), 81–112.

Deyhle, D., & Swisher, K. (1997). Research in American Indian and Alaska Native education: From assimilation to self-determination. In M. W. Apple (ed.), *Review of Research in Education*, Vol. 22 (pp. 113–194). Washington, DC: American Educational Research Association.

diSessa, A., Hammer, D., Sherin, B., & Kolpakowski, T. (1991). Inventing graphing: Meta-representational expertise in children. *Journal of Mathematical Behavior*, 10(2), 117–160.

Dixon, J. K. (1995). Limited English proficiency and spatial visualization in middle school students' construction of the concepts of reflection and rotation. *Bilingual Research Journal*, 19(2), 221–247.

Driver, R., Asoko, H., Leach, J., Mortimer, E., & Scott, P. (1994). Constructing scientific knowledge in the classroom. *Educational Researcher*, 23(7), 5–12.

Duran, B. J., Dugan, T., & Weffer, R. (1998). Language minority students in high school: The role of language in learning biology concepts. *Science Education*, 82(3), 311–341.

Dzama, E. N. N., & Osborne, J. F. (1999). Poor performance in science among African students: An alternative explanation to the African worldview thesis. *Journal of Research in Science Teaching*, 36(3), 387–405.

Eide, K. Y., & Heikkinen, M. W. (1998). The inclusion of multicultural material in middle school science teachers' resource manuals. *Science Education*, 82(2), 181–195.

Eisenhart, M., Finkel, E., & Marion, S. F. (1996). Creating the conditions for scientific literacy: A re-examination. *American Educational Research Journal*, 33(2), 261–295.

Elmore, R. (1996). Getting to scale with good educational practice. *Harvard Educational Review*, 66(1), 1–26.

Epstein, J. (1987). Parent involvement: What research says to administrators. *Education and Urban Society*, 19(2), 119–136.

Fishman, B., Marx, R. W., Blumenfeld, P., Krajcik, J., & Soloway, E. (2004). Creating a framework for research on systemic technology innovations. *Journal of the Learning Sciences*, 13(1), 43–76.

Fradd, S. H., & Lee, O. (1995). Science for all: A promise or a pipe dream for bilingual students? *Bilingual Research Journal*, 19(2), 261–278.

Fradd, S. H., & Lee, O. (1999). Teachers' roles in promoting science inquiry with students from diverse language backgrounds. *Educational Researcher*, 28(6), 4–20, 42.

Fradd, S. H., & Lee, O. (2000). Needed: A framework for integrating standardized and informal assessment for students developing academic language proficiency in English. In J. V. Tenajero & S. Hurley (eds.), *Literacy assessment of bilingual learners* (pp. 130–148). Boston: Allyn and Bacon.

Fradd, S. H., Lee, O., Sutman, F. X., & Saxton, M. K. (2002). Materials development promoting science inquiry with English language learners: A case study. *Bilingual Research Journal*, 25(4), 479–501.

Fusco, D. (2001). Creating relevant science through urban planning and gardening. *Journal of Research in Science Teaching*, 38(8), 860–877.

Fusco, D., & Calabrese Barton, A. (2001). Representing student achievement in science. *Journal of Research in Science Teaching*, 38(3), 337–354.

Gamoran, A., Anderson, C. W., Quiroz, P. A., Secada, W. G., Williams, T., & Ashmann, S. (2003). *Transforming teaching in math and science: How schools and districts can support change*. New York: Teachers College Press.

Gao, L. (1998). Cultural context of school science teaching and learning in the People's Republic of China. *Science Education*, 82(1), 1–13.

Gao, L., & Watkins, D. A. (2002). Conceptions of teaching held by school science teachers in P. R. China: Identification and cross-cultural comparisons. *International Journal of Science Education*, 24(1), 61–79.

Garaway, G. B. (1994). Language, culture, and attitude in mathematics and science learning: A review of the literature. *Journal of Research and Development in Education*, 27(2), 102–111.

García, E. E. (1999). *Student cultural diversity: Understanding and meeting the challenge*. 2nd ed. Boston: Houghton Mifflin.

García, E. E., & Curry Rodríguez, J. (2000). The education of limited English proficient students in California schools. *Bilingual Research Journal*, 24(1–3), 15–35.

Garet, M. S., Porter, A. C., Desimone, L., Birman, B. F., & Yoon, K. S. (2001). What makes professional development effective? Results from a national sample of teachers. *American Educational Research Journal*, 38(4), 915–945.

Gay, G. (2002). Preparing for culturally responsive teaching. *Journal of Teacher Education*, 53(2), 106–116.

George, J. (1992). Science teachers as innovators using indigenous resources. *International Journal of Science Education*, 14(1), 95–109.

George, J. (1999). Worldview analysis of knowledge in a rural village: Implications for science education. *Science Education*, 83(1), 77–95.

George, Y. S., Neale, D. S., Van Horne, V., & Malcom, S. M. (2001). *In pursuit of a diverse science, technology, engineering, and mathematics workforce: Recommended research priorities to enhance participation by underrepresented minorities*. Washington, DC: American Association for the Advancement of Science.

Gilbert, A., & Yerrick, R. (2001). Same school, separate worlds: A sociocultural study of identity, resistance, and negotiation in a rural, lower track science classroom. *Journal of Research in Science Teaching*, 38(5), 574–598.

Giroux, H. (1992). *Border crossings: Cultural workers and the politics of education*. New York: Routledge.

Grandy, J. (1998). Persistence in science of high-ability minority students: Results of a longitudinal study. *Journal of Higher Education*, 69(6), 589–620.

Gutiérrez, K. D., Asato, J., Pacheco, M., Moll, L. C., Olson, K., Horng, E. L., Ruiz, R., García, E., & McCarty, T. (2002). "Sounding American": The consequences of new reforms on English language learners. *Reading Research Quarterly*, 37(3), 328–343.

Gutiérrez, K. D., & Rogoff, B. (2003). Cultural ways of learning: Individual traits of repertoires of practice. *Educational Researcher*, 32(5), 19–25.

Haberman, M. (1988). Proposals for recruiting minority teachers: Promising practices and attractive detours. *Journal of Teacher Education*, 33(4), 38–44.

Hamilton, L. S., Nussbaum, E. M., Kupermintz, H., Kerkhoven, J. I. M., & Snow, R. E. (1995). Enhancing the validity and usefulness of large-scale educational assessments: II. NELS: 88 science achievement. *American Educational Research Journal, 32*(3), 555–581.

Hammond, L. (2001). An anthropological approach to urban science education for language minority families. *Journal of Research in Science Teaching, 38*(9), 983–999.

Hampton, E., & Rodriguez, R. (2001). Inquiry science in bilingual classrooms. *Bilingual Research Journal, 25*(4), 461–478.

Haraway, D. J. (1990). *Primate visions: Gender, race and nature in the world of modern science.* New York: Routledge.

Haraway, D. J. (1991). *Simians, cyborgs, and women: The reinvention of nature.* New York: Routledge.

Hart, J., & Lee, O. (2003). Teacher professional development to improve science and literacy achievement of English language learners. *Bilingual Research Journal, 27*(3), 475–501.

Haycock, K. (1999). *Dispelling the myth: High poverty schools exceeding expectations.* Washington, DC: Education Trust.

Hayes, M. T., & Deyhle, D. (2001). Constructing difference: A comparative study of elementary science curriculum differentiation. *Science Education, 85*(3), 239–262.

Heath, S. B. (1983). *Ways with words: Language, life, and work in communities and classroom.* New York: Cambridge University Press.

Hewson, M. G. (1988). The ecological context of knowledge: Implications for learning science in developing countries. *Journal of Curriculum Studies, 20*(4), 317–326.

Hewson, P. W., Kahle, J. B., Scantlebury, K., & Davies, D. (2001). Equitable science education in urban middle schools: Do reform efforts make a difference? *Journal of Research in Science Teaching, 38*(10), 1130–1144.

Hill, O. W., Pettus, C., & Hedin, B. A. (1990). Three studies of factors affecting the attitudes of blacks and females toward the pursuit. *Journal of Research in Science Teaching, 27*(4), 289–314.

Hilliard, A. G. (2003). No mystery: Closing the achievement gap between Africans and excellence. In T. Perry, C. Steele, & A. G. Hilliard (eds.), *Young, gifted, and Black: Promoting high achievement among African American students* (pp. 131–165). Boston: Beacon Press.

Hodson, D. (1993). In search of a rationale for multicultural science education. *Science Education, 77*(6), 685–711.

Hodson, D. (1999). Going beyond cultural pluralism: Science education for sociopolitical action. *Science Education, 83*(6), 775–796.

Hodson, D., & Dennick, R. (1994). Antiracist education: A special role for the history of science and technology. *School Science and Mathematics, 94*(5), 255–262.

Howes, E. V. (2002). Learning to teach science for all in the elementary grades: What do prospective teachers bring? *Journal of Research in Science Teaching, 39*(9), 845–869.

Irzik, G., & Irzik, S. (2002). Which multiculturalism? *Science & Education, 11*(4), 393–403.

Jegede, O. J., & Aikenhead, G. S. (1999). Transcending cultural borders: Implications for science teaching. *Research in Science and Technology Education, 17*(1), 45–66.

Jegede, O. J., & Okebukola, P. A. (1991a). The effect of instruction on socio-cultural beliefs hindering the learning of science. *Journal of Research in Science Teaching, 28*(3), 275–285.

Jegede, O. J., & Okebukola, P. A. (1991b). The relationship between African traditional cosmology and students' acquisition of a science process skill. *International Journal of Science Education,* 13(1), 37–47.

Jegede, O. J., & Okebukola, P. A. (1992). Differences in sociocultural environment perceptions associated with gender in science classrooms. *Journal of Research in Science Teaching,* 29(7), 637–647.

Jiménez, R. T., & Gersten, R. (1999). Lessons and dilemmas derived from the literacy instruction of two Latina/o teachers. *American Educational Research Journal,* 36(2), 265–301.

Johnson, J., & Kean, E. (1992). Improving science teaching in multicultural settings: A qualitative study. *Journal of Science Education and Technology,* 1(4), 275–287.

Jorgenson, O. (2000). The need for more ethnic teachers: Addressing the critical shortage in American public schools. Teachers College Record. Online format only. http://www.tcrecord.org. ID Number: 10551. Published: 9/13/2000.

Kahle, J. B. (1982). Can positive minority attitudes lead to achievement gains in science? Analysis of the 1977 National Assessment of Educational Progress, attitudes toward science. *Science Education,* 66(4), 539–546.

Kahle, J. B. (1998). Equitable systemic reform in science and mathematics: Assessing progress. *Journal of Women and Minorities in Science and Engineering,* 4(2–3), 91–112.

Kahle, J. B., & Kelly, M. K. (2001). Equity in reform: Case studies of five middle schools involved in systemic reform. *Journal of Women and Minorities in Science and Engineering,* 7(2), 79–96.

Kahle, J. B., Meece, J., & Scantlebury, K. (2000). Urban African-American middle school science students: Does standards-based teaching make a difference? *Journal of Research in Science Teaching,* 37(9), 1019–1041.

Kawagley, A. O., Norris-Tull, D., & Norris-Tull, R. A. (1998). The indigenous worldview of Yupiaq culture: Its scientific nature and relevance to the practice and teaching of science. *Journal of Research in Science Teaching,* 35(2), 133–144.

Kawasaki, K. (1996). The concepts of science in Japanese and Western education. *Science & Education,* 5(1), 1–20.

Kearsey, J., & Turner, S. (1999). The value of bilingualism in pupils' understanding of scientific language. *International Journal of Science Education,* 21(10), 1037–1050.

Keller, E. F., & Longino, H. E. (eds.). (1996). *Feminism and science.* New York: Oxford University Press.

Kelly, G. J., & Breton, T. (2001). Framing science as disciplinary inquiry in bilingual classrooms. *Electronic Journal of Literacy Through Science,* 1(1). http://www2.sjsu.edu/elementaryed/ejlts/. Accessed October 31, 2002.

Kelly, G. J., Carlsen, W. S., & Cunningham, C. M. (1993). Science education in sociocultural context: Perspectives from the sociology of science. *Science Education,* 77(2), 207–220.

Kennedy, M. (1998). *Form and substance in inservice teacher education* (Research Monograph No. 13). Madison: University of Wisconsin, National Institute for Science Education.

Kesamang, M. E. E., & Taiwo, A. A. (2002). The correlates of the socio-cultural background of Botswana junior secondary school students with their attitudes towards and achievements in science. *International Journal of Science Education,* 24(9), 919–931.

Kim, J. J., Crasco, L., Smith, R. B., Johnson, G., Karantonis, A., & Leavitt, D. J. (2001a). *Academic excellence for all students: Their accomplishment in science and mathematics.* Norwood, MA, Systemic Research, Inc.

Kim, J. J., Crasco, L. M., Smithson, J., & Blank, R. K. (2001b). *Survey results of urban school classroom practices in mathematics and science: 2000 report.* Norwood, MA: Systemic Research, Inc.

King, K., Shumow, L., & Lietz, S. (2001). Science education in an urban elementary school: Case studies of teacher beliefs and classroom practices. *Science Education,* 85(2), 89–110.

Klein, C. A. (1982). Children's concepts of the earth and the sun: A cross cultural study. *Science Education,* 66(1), 95–107.

Klein, S. P., Jovanovic, J., Stecher, B. M., McCaffrey, D., Shavelson, R. J., Haertel, E., Solano-Flores, G., & Comfort, K. (1997). Gender and racial/ethnic differences on performance assessment in science. *Educational Evaluation and Policy Analysis,* 19(2), 83–97.

Knapp, M. S. (1997). Between systemic reforms and the mathematics and science classroom: The dynamics of innovation, implementation, and professional learning. *Review of Educational Research,* 67(2), 227–266.

Knapp, M. S., & Plecki, M. L. (2001). Investing in the renewal of urban science teaching. *Journal of Research in Science Teaching,* 38(10), 1089–1100.

Krugly-Smolska, E. (1996). Scientific culture, multiculturalism and the science classroom. *Science & Education,* 5(1), 21–29.

Kuhn, T. S. ([1970] 1996). *The structure of scientific revolutions.* Chicago: University of Chicago Press.

Kurth, L. A., Anderson, C. W., & Palincsar, A. S. (2002). The case of Carla: Dilemmas of helping *all* students to understand science. *Science Education,* 86(3), 287–313.

Labov, W. (1966). *The social stratification of English in New York City.* Washington, DC: Center for Applied Linguistics.

Lacelle-Peterson, M. W., & Rivera, C. (1994). Is it real for all kids? A framework for equitable assessment policies for English language learners. *Harvard Educational Review,* 64(1), 55–75.

Ladson-Billings, G. (1994). *The dreamkeepers: Successful teachers of African American children.* San Francisco: Jossey-Bass.

Ladson-Billings, G. (1995). Toward a theory of culturally relevant pedagogy. *American Educational Research Journal,* 32(3), 465–491.

Ladson-Billings, G. (1999). Preparing teachers for diverse student populations: A critical race theory perspective. In A. Iran-Nejad & P. D. Pearson (eds.), *Review of Research in Education.* Vol. 24 (pp. 211–248). Washington, DC: American Educational Research Association.

Lambert, J., Lester, B., Lee, O., & Luykx, A. (in press). Changing teachers' beliefs about science and student diversity through an inquiry-based earth systems curricular and professional development intervention. *Journal of Science Teacher Education.*

Latour, B., & Woolgar, S. (1986). *Laboratory life: The social construction of scientific facts.* Princeton, NJ: Princeton University Press.

Lawrenz, F., & Gray, B. (1995). Investigation of worldview theory in a South African context. *Journal of Research in Science Teaching,* 32(6), 555–568.

Lawrenz, F., Huffman, D., & Welch, W. (2001). The science achievement of various subgroups of alternative assessment formats. *Science Education*, 85(3), 279–290.

Lee, C. D. (2001). Is October Brown Chinese? A cultural modeling activity system for underachieving students. *American Educational Research Journal*, 38(1), 97–141.

Lee, H.-S., & Songer, N. B. (2003). Making authentic science accessible to students. *International Journal of Science Education*, 25(1), 1–26.

Lee, O. (1996). Diversity and equity for Asian American students in science education. *Science Education*, 81(1), 107–122.

Lee, O. (1999a). Equity implications based on the conceptions of science achievement in major reform documents. *Review of Educational Research*, 69(1), 83–115.

Lee, O. (1999b). Science knowledge, worldviews, and information sources in social and cultural contexts: Making sense after a natural disaster. *American Educational Research Journal*, 36(2), 187–219.

Lee, O. (2002). Science inquiry for elementary students from diverse backgrounds. In W. G. Secada (ed.), *Review of Research in Education*. Vol. 26 (pp. 23–69). Washington, DC: American Educational Research Association.

Lee, O. (2003). Equity for culturally and linguistically diverse students in science education: A research agenda. *Teachers College Record*, 105(3), 465–489.

Lee, O. (2004). Teacher change in beliefs and practices in science and literacy instruction with English language learners. *Journal of Research in Science Teaching*, 41(1), 65–93.

Lee, O., & Avalos, M. (2002). Promoting science instruction and assessment for English language learners. *Electronic Journal of Science Education*, 7(2). http://unr.edu/homepage/crowther/ejse/. Accessed April 5, 2003.

Lee, O., Deaktor, R. A., Hart, J. E., Cuevas, P., & Enders, C. (2005). An instructional intervention's impact on the science and literacy achievement of culturally and linguistically diverse elementary students. *Journal of Research in Science Teaching*, 42(8), 857–887.

Lee, O., & Fradd, S. H. (1996a). Literacy skills in science performance among culturally and linguistically diverse students. *Science Education*, 80(6), 651–671.

Lee, O., & Fradd, S. H. (1996b). Interactional patterns of linguistically diverse students and teachers: Insights for promoting science learning. *Linguistics and Education: An International Research Journal*, 8(2), 269–297.

Lee, O., & Fradd, S. H. (1998). Science for all, including students from non-English language backgrounds. *Educational Researcher*, 27(3), 12–21.

Lee, O., Fradd, S. H., & Sutman, F. X. (1995). Science knowledge and cognitive strategy use among culturally and linguistically diverse students. *Journal of Research in Science Teaching*, 32(8), 797–816.

Lee, O., Hart, J., Cuevas, P., & Enders, C. (2004). Professional development in inquiry-based science for elementary teachers of diverse students. *Journal of Research in Science Teaching*, 41(10), 1021–1043.

Lee, O., & Luykx, A. (2005). Dilemmas in scaling up educational innovations with nonmainstream students in elementary school science. *American Educational Research Journal*, 42(3), 411–438.

Lee, O., & Paik, S. (2000). Conceptions of science achievement in major reform documents. *School Science and Mathematics*, 100(1), 16–26.

Lee, V., & Smith, J. B. (1993). Effects of school restructuring on the achievement and engagement of middle grade students. *Sociology of Education*, 66(3), 164–187.

Lee, V., & Smith, J. B. (1995). Effects of high school restructuring and size on gains in achievement and engagement for early secondary school students. *Sociology of Education*, 68(4), 241–247.

Lee, V., Smith, J., Croninger, J. B., & Robert, G. (1997). How high school organization influences the equitable distribution of learning in mathematics and science. *Sociology of Education*, 70(2), 128–150.

Lehrer, R., & Schauble, L. (2000). Modeling in mathematics and science. In R. Glaser (ed.), *Advances in instructional psychology*. Vol. 5. Mahwah, NJ: Lawrence Erlbaum.

Lemke, J. L. (1990). *Talking science: Language, learning and values*. Norwood, NJ: Ablex.

Lemke, J. L. (2001). Articulating communities: Sociocultural perspectives on science education. *Journal of Research in Science Teaching*, 38(3), 296–316.

Lemmer, M., Lemmer, T. N., & Smit, J. J. A. (2003). South African students' views of the universe. *International Journal of Science Education*, 25(5), 563–582.

Levinson, B. A., Foley, D. E., & Holland, D. C. (eds.). (1996). *The cultural production of the educated person: Critical ethnographies of schooling and local practice*. Albany: State University of New York Press.

Lipka, J. (1998). *Transforming the culture of schools: Yup'ik Eskimo examples*. Mahwah, NJ: Erlbaum.

Lipka, J., & Adams, B. (2004). *Culturally based math education as a way to improve Alaska Native students' math performance*. Athens: Ohio University, Appalachian Collaborative Center for Learning, Assessment, and Instruction in Mathematics.

Loucks-Horsley, S., Hewson, P. W., Love, N., & Stiles, K. E. (1998). *Designing professional development for teachers of science and mathematics*. Thousand Oaks, CA: Corwin.

Loving, C. C. (1997). From the summit of truth to its slippery slopes: Science education's journey through positivist-postmodern territory. *American Educational Research Journal*, 34(3), 421–452.

Loving, C. C. (1998). Cortes' multicultural empowerment model and generative teaching and learning in science. *Science & Education*, 7(6), 533–552.

Loving, C. C., & Marshall, J. E. (1997). Increasing the pool of ethnically diverse science teachers: A mid-project evaluation. *Journal of Science Teacher Education*, 8(3), 205–217.

Lubienski, S. (2003). Celebrating diversity and denying disparities: A critical assessment. *Educational Researcher*, 32(8), 30–38.

Luft, J. A. (1999). The border crossings of a multicultural science education enthusiast. *School Science and Mathematics*, 99(7), 380–388.

Luft, J. A., Bragg, J., & Peters, C. (1999). Learning to teach in a diverse setting: A case study of a multicultural science education enthusiast. *Science Education*, 83(5), 527–543.

Luykx, A., Cuevas, P., Lambert, J., & Lee, O. (2005). Unpacking teachers' "resistance" to integrating students' language and culture into elementary science instruction. In A. Rodriguez & R. S. Kitchen (eds.), *Preparing mathematics and science teachers for diverse classrooms: Promising strategies for transformative pedagogy* (pp. 119–141). Mahwah, NJ: Lawrence Erlbaum.

Luykx, A., & Lee, O. (in press). Measuring instructional congruence in elementary science classrooms: Pedagogical and methodological components of a theoretical framework. *Journal of Research in Science Teaching*.

Luykx, A., Lee, O., & Edwards, U. (in press). Lost in translation: Negotiating meanings in a beginning ESOL science classroom. *Educational Policy*.

Luykx, A., Lee, O., Mahotiere, M., Lester, B., Hart, J., & Deaktor, R. (in press). Cultural and home language influence in elementary students' constructed responses on science assessments. *Teachers College Record*.

Lynch, M. (1985). *Art and artifact in laboratory science: A study of shop work and shop talk in a research laboratory.* Boston: Routledge and Kegan Paul.

Lynch, P. P. (1996a). Students' alternative frameworks: Linguistic and cultural interpretations based on a study of a western-tribal continuum. *International Journal of Science Education*, 18(3), 321–332.

Lynch, P. P. (1996b). Students' alternative frameworks for the nature of matter: A cross-cultural study of linguistic and cultural interpretations. *International Journal of Science Education*, 18(6), 743–752.

Lynch, P. P., Chipman, H. H., & Pachaury, A. C. (1985a). The language of science and the high school student: The recognition of concept definitions: A comparison between Hindi speaking students in India and English speaking students in Australia. *Journal of Research in Science Teaching*, 22(7), 675–686.

Lynch, P. P., Chipman, H. H., & Pachaury, A. C. (1985b). The language of science and preferential thinking styles: A comparison between Hindi speaking students (in India) and English speaking students (in Australia). *Journal of Research in Science Teaching*, 22(8), 699–712.

Lynch, S. (2000). *Equity and science education reform.* Mahwah, NJ: Erlbaum.

Lynch, S., Kuipers, J., Pyke, C., & Szesze, M. (2005). Examining the effects of a highly rated science curriculum unit on diverse populations: Results from a planning grant. *Journal of Research in Science Teaching*, 42(8), 912–946.

Madaus, G. F. (1994). A technological and historical consideration of equity issues associated with proposals to change the nation's testing policy. *Harvard Educational Review*, 64(1), 76–95.

Maple, S., & Stage, F. (1991). Influences on the choice of math/science major by gender and ethnicity. *American Educational Research Journal*, 28(1), 37–60.

Mark, R. W., Blumenfeld, P. C., Krajcik, J. S., Fishman, B., Soloway, E., Geier, R., & Tal, R. T. (2004). Inquiry-based science in the middle grades: Assessment of learning in urban systemic reform. *Journal of Research in Science Teaching*, 41(10), 1063–1080.

Martin, M. O., Mullis, I. V. S., Gonzalez, E. J., O'Connor, K. M., Chrostowski, S. J., Gregory, K. D., Smith, T. A., & Garden, R. A. (2001). *Science benchmarking report TIMSS 1999-eighth grade: Achievement for U.S. states and districts in an international context.* Chestnut Hill, MA: Boston College, The International Study Center.

Matthews, C. E., & Smith, W. S. (1994). Native American related materials in elementary science instruction. *Journal of Research in Science Teaching*, 31(4), 363–380.

McCarty, T. L., Lynch, R. H., Wallace, S., & Benally, A. (1991). Classroom inquiry and Navajo learning styles: A call for reassessment. *Anthropology and Education Quarterly*, 22(1), 42–59.

McKinley, E. (2004). Locating the global: Culture, language and science education for indigenous students. *International Journal of Science Education*, 27(2), 227–241.

McKinley, E., Waiti, P. M., & Bell, B. (1992). Language, culture and science education. *International Journal of Science Education*, 14(5), 579–595.

McLaughlin, M. W., Shepard, L. A., & O'Day, J. A. (1995). *Improving education through standards-based reform: A report by the National Academy of Education Panel on Standards-based Education Reform.* Stanford, CA: Stanford University, National Academy of Education.

McNeil, L. M. (2000). Creating new inequalities: Contradictions of reform. *Phi Delta Kappan*, 81(10), 729–734.

Merino, B., & Hammond, L. (2001). How do teachers facilitate writing for bilingual learners in "sheltered constructivist" science? *Electronic Journal of Literacy through Science*, 1(1). http://www2.sjsu.edu/elementaryed/ejlts/. Accessed October 31, 2002.

Metz, K. E. (1998). Scientific inquiry within reach of young children. In B. J. Fraser, & K. Tobin (eds.), *International handbook of science education.* Part I (pp. 81–96). Dordrecht, the Netherlands: Kluwer Academic Publishers.

Moje, E., Collazo, T., Carillo, R., & Marx, R. W. (2001). "Maestro, what is quality?": Examining competing discourses in project-based science. *Journal of Research in Science Teaching*, 38(4), 469–495.

Moll, L. C. (1992). Bilingual classroom studies and community analysis: Some recent trends. *Educational Researcher*, 21(2), 20–24.

Muller, P. A., Stage, F. K., & Kinzie, J. (2001). Science achievement growth trajectories: Understanding factors related to gender and racial-ethnic differences in precollege science achievement. *American Educational Research Journal*, 38(4), 981–1012.

Murfin, B. (1994). African science, African and African-American scientists and the school science curriculum. *School Science and Mathematics*, 94(2), 96–103.

National Center for Children in Poverty. (1995). *Five million children: A statistical profile of our poorest young citizens.* New York: Columbia University Press.

National Center for Education Statistics. (1996). *Pursuing excellence: A study of U.S. eighth-grade mathematics and science teaching, learning, curriculum, and achievement in international context.* Washington, DC: U.S. Department of Education, Office of Educational Research and Improvement.

National Center for Education Statistics. (1997). *The condition of education, 1997.* Washington, DC: U.S. Department of Education.

National Center for Education Statistics. (1999). *Teacher quality: A report on the preparation and qualifications of public school teachers.* Washington, DC: U. S. Department of Education, Office of Educational Research and Improvement.

National Research Council (NRC). (1996). *National science education standards.* Washington, DC: National Academy Press.

National Research Council (NRC). (2000). *Inquiry and the national science education standards: A guide for teaching and learning.* Washington, DC: National Academy Press.

National Science Foundation (NSF). (1996). *Review of instructional materials for middle school science.* Washington, DC: National Science Foundation.

National Science Foundation (NSF). (1998). *Infusing equity in systemic reform: An implementation scheme.* Washington, DC: National Science Foundation.

National Science Foundation (NSF). (2002). *Women, minorities, and persons with disabilities in science and engineering.* Arlington, VA: National Science Foundation.

Nelson-Barber, S., & Estrin, E. T. (1995). Bringing Native American perspectives to mathematics and science teaching. *Theory into Practice,* 34(3), 174–185.

Nelson-Barber, S., & Estrin, E. T. (1996). *Culturally responsive mathematics and science education for Native students.* San Francisco: Far West Laboratory for Educational Research and Development.

Ninnes, P. (1994). Toward a functional learning system for Solomon Island secondary science classrooms. *International Journal of Science Education,* 16(6), 677–688.

Ninnes, P. (1995). Informal learning contexts in Solomon Islands and their implications for the cross-cultural classroom. *International Journal of Educational Development,* 15(1), 15–26.

Ninnes, P. (2000). Representations of indigenous knowledges in secondary school science textbooks in Australia and Canada. *International Journal of Science Education,* 22(6), 603–617.

No Child Left Behind Act of 2001. Public Law No. 107–110, 115 Stat. 1425. (2002).

Norman, O. (1998). Marginalized discourses and scientific literacy. *Journal of Research in Science Teaching,* 35(4), 365–374.

Norman, O., Ault, C. R., Bentz, B., & Meskimen, L. (2001). The Black-White "achievement gap" as a perennial challenge of urban science education: A sociocultural and historical overview with implications for research and practice. *Journal of Research in Science Teaching,* 38(10), 1101–1114.

Nussbaum, E. M., Hamilton, L. S., & Snow, R. E. (1997). Enhancing the validity and usefulness of large-scale educational assessments: IV. NELS: 88 science achievement to 12th grade. *American Educational Research Journal,* 34(1), 151–173.

Oakes, J. (1990). Opportunities, achievement, and choice: Women and minority students in science and mathematics. In C. B. Cazden (ed.), *Review of Research in Education.* Vol. 16 (pp. 153–221). Washington, DC: American Educational Research Association.

Ochs, E., Jacoby, S., & Gonzales, P. (1996). "When I come down I'm in the domain state": Grammar and graphic representation in the interpretive activity of physicists. In E. Ochs, E. A. Schegloff, & S. A. Thompson (eds.), *Interaction and grammar* (pp. 328–369). New York: Cambridge University Press.

Ogawa, M. (1995). Science education in a multiscience perspective. *Science Education,* 79(5), 583–593.

Ogbu, J. (1999). Beyond language: Ebonics, proper English, and identity in a Black-American speech community. *American Educational Research Journal,* 36(2), 147–184.

Ogbu, J., & Simons, H. D. (1998). Voluntary and involuntary minorities: A cultural-ecological theory of school performance with some implications for education. *Anthropology and Education Quarterly,* 29(2), 155–188.

Ogunniyi, M. B. (1987). Conceptions of traditional cosmological ideas among literate and nonliterate Nigerians. *Journal of Research in Science Teaching*, 24(2), 107–117.

Ogunniyi, M. B. (1988). Adapting Western science to traditional African culture. *International Journal of Science Education*, 10(1), 1–9.

Ogunniyi, M. B., Jegede, O. J., Ogawa, M., Yandila, C. D., & Oladele, F. K. (1995). Nature of worldview presuppositions among science teachers in Botswana, Indonesia, Japan, Nigeria, and the Philippines. *Journal of Research in Science Teaching*, 32(8), 817–832.

Okebukola, P. A., & Jegede, O. J. (1990). Eco-cultural influences upon students' concept attainment in science. *Journal of Research in Science Teaching*, 27(7), 651–660.

O'Loughlin, M. (1992). Rethinking science education: Beyond Piagetian constructivism toward a sociocultural model of teaching and learning. *Journal of Research in Science Teaching*, 29(8), 791–820.

Osborne, A. B. (1996). Practice into theory into practice: Culturally relevant pedagogy for students we have marginalized and normalized. *Anthropology and Education Quarterly*, 27(3), 285–314.

O'Sullivan, C. Y., Lauko, M. A., Grigg, W. S., Qian, J., & Zhang, J. (2003). *The nation's report card: Science 2000.* Washington, DC: U.S. Department of Education, Institute of Education Sciences.

Peng, S., & Hill, S. (1994). Characteristics and educational experiences of high-achieving minority secondary students in science and mathematics. *Journal of Women and Minorities in Science and Engineering*, 1, 137–152.

Pew Charitable Trust. (1998). Quality counts '98: The urban challenge: Public education in the 50 states. *Education Week on the Web*, January 8, http://counts.edweek.org/sreports/qc98/qc98to.htm.

Porter, A. C. (1995). The uses and misuses of opportunity-to-learn standards. *Educational Researcher*, 24(1), 21–27.

Powell, R. R., & Garcia, J. (1985). The portrayal of minorities and women in selected elementary science series. *Journal of Research in Science Teaching*, 22(6), 519–533.

Prophet, R. B. (1990). Rhetoric and reality in science curriculum development in Botswana. *International Journal of Science Education*, 12(1), 13–23.

Prophet, R. B., & Powell, P. M. (1993). Coping and control: Science teaching strategies in Botswana. *Qualitative Studies in Education*, 6, 197–209.

Rahm, J. (2002). Emergent learning opportunities in an inner-city youth gardening program. *Journal of Research in Science Teaching*, 39(2), 164–184.

Raizen, S. (1998). Standards for science education. *Teachers College Record*, 100(1), 66–121.

Rakow, S. J. (1985a). Minority students in science: Perspectives from the 1981–1982 National Assessment in Science. *Urban Education*, 20(1), 103–113.

Rakow, S. J. (1985b). Prediction of the science inquiry skill of seventeen-year-olds: A test of the model of educational productivity. *Journal of Research in Science Teaching*, 22(4), 289–302.

Rakow, S. J., & Bermudez, A. B. (1993). Science is "Cicencia": Meeting the needs of Hispanic American students. *Science Education*, 77(6), 547–560.

Raudenbush, S. W. (2003). *Designing field trials of educational innovations.* Paper presented at the conference, "Conceptualizing Scale-Up: Multidisciplinary Perspectives," Washington, DC, November.

Rennie, L. J. (1998). Gender equity: Toward clarification and a research direction for science teacher education. *Journal of Research in Science Teaching*, 35(8), 951–961.

Reyes, M. (1992). Challenging venerable assumptions: Literacy instruction for linguistically diverse students. *Harvard Educational Review*, 62(4), 427–446.

Richardson, V., & Placier, P. (2001). Teacher change. In D. V. Richardson (ed.), *Handbook of research on teaching*. 4th ed. (pp. 905–950). Washington, DC: American Educational Research Association.

Riggs, E. M. (2005). Field-based education and indigenous knowledge: Essential components of geoscience education for Native American communities. *Science Education*, 89(2), 296–313.

Rivet, A. E., & Krajcik, J. S. (2004). Achieving standards in urban systemic reform: An example of a sixth grade project-based science curriculum. *Journal of Research in Science Teaching*, 41(7), 669–693.

Rodriguez, A. (1997). The dangerous discourse of invisibility: A critique of the NRC's National Science Education Standards. *Journal of Research in Science Teaching*, 34(1), 19–37.

Rodriguez, A. (1998a). Busting open the meritocracy myth: Rethinking equity and student achievement in science education. *Journal of Women and Minorities in Science and Engineering*, 4(2–3), 195–216.

Rodriguez, A. (1998b). Strategies for counter-resistance: Toward sociotransformative constructivism and learning to teach science for diversity and for understanding. *Journal of Research in Science Teaching*, 35(6), 589–622.

Rodriguez, A. J. (2001). From gap gazing to promising cases: Moving toward equity in urban education reform. *Journal of Research in Science Teaching*, 38(10), 1115–1129.

Rodriguez, A. J., & Berryman, C. (2002). Using sociotransformative constructivism to teach for understanding in diverse classrooms: A beginning teacher's journey. *American Educational Research Journal*, 39(4), 1017–1045.

Rodriguez, I., & Bethel, L. J. (1983). An inquiry approach to science and language teaching. *Journal of Research in Science Teaching*, 20(4), 291–296.

Rollnick, M., & Rutherford, M. (1996). The use of mother tongue and English in the learning and expression of science concepts: A classroom-based study. *International Journal of Science Education*, 18(1), 91–103.

Rosebery, A. S., Warren, B., & Conant, F. R. (1992). Appropriating scientific discourse: Findings from language minority classrooms. *Journal of the Learning Sciences*, 21(1), 61–94.

Roth, W.-M., Tobin, K., Carambo, C., & Dalland, C. (2004). Coteaching: Creating resources for learning and learning to teach chemistry in urban high schools. *Journal of Research in Science Teaching*, 41(9), 882–904.

Ruiz-Primo, M. A., & Shavelson, R. J. (1996). Rhetoric and reality in science performance assessments: An update. *Journal of Research in Science Teaching*, 33(10), 1045–1063.

Schibeci, R. A., & Riley, J. P. (1986). Influence of students' background and perceptions of science attitudes and achievement. *Journal of Research in Science Teaching*, 23(3), 177–187.

Schmidt, W. H., McKnight, C. C., & Raizen, S. A. (1997). *A splintered vision: An investigation of U.S. science and mathematics education.* Dordrecht, the Netherlands: Kluwer Academic Publishers.

Sconiers, Z. D., & Rosiek, J. L. (2000). Historical perspective as an important element of teachers' knowledge: A sonata-form case study of equity issues in a chemistry classroom: Voices inside schools. *Harvard Educational Review,* 70(3), 370–404.

Secada, W. G., & Lee, O. (2003). *A study of highly effective USI schools in the teaching of mathematics and science: Classroom level results.* Washington, DC: The Urban Institute.

Seiler, G. (2001). Reversing the "standard" direction: Science emerging from the lives of African American students. *Journal of Research in Science Teaching,* 38(9), 1000–1014.

Setati, M., Adler, J., Reed, Y., & Bapoo, A. (2002). Incomplete journeys: Code-switching and other language practices in mathematics, science and English language classrooms in South Africa. *Language and Education,* 16(2), 128–150.

Settlage, J., & Meadows, L. (2002). Standards-based reform and its unintended consequences: Implications for science education within America's urban schools. *Journal of Research in Science Teaching,* 39(2), 114–127.

Shavelson, R. J., & Towne, L. (eds.). (2002). *Scientific research in education.* Washington, DC: National Academy Press.

Shaw, J. M. (1997). Threats to the validity of science performance assessments for English language learners. *Journal of Research in Science Teaching,* 34(7), 721–743.

Shumba, O. (1999). Relationship between secondary science teachers' orientation to traditional culture and beliefs concerning science instructional ideology. *Journal of Research in Science Teaching,* 36(3), 333–355.

Siegel, H. (1995). What price inclusion? *Teachers College Record,* 97(1), 6–31.

Siegel, H. (1997). Science education: Multicultural and universal. *Interchange,* 28(2), 97–108.

Siegel, H. (2002). Multiculturalism, universalism, and science education: In search of common ground. *Science Education,* 86(6), 803–820.

Sizemore, B., Brosard, C., & Harrigan, B. (1994). *An abashing anomaly: The high achieving predominantly Black elementary schools.* Pittsburgh, PA: University of Pittsburgh Press.

Smith, F. M., & Hausafus, C. O. (1998). Relationship of family support and ethnic minority students' achievement in science and mathematics. *Science Education,* 82(1), 111–125.

Smith, M. S., & O'Day, J. (1991). Systemic school reform. In S. H. Fuhrman & B. Malen (eds.), *The politics of curriculum and testing: The 1990 yearbook of the Politics of Education Association* (pp. 233–268). Briston, PA: Falmer Press.

Snively, G. (1990). Traditional Native Indian beliefs, cultural values, and science instruction. *Canadian Journal of Native Education,* 17(1), 44–59.

Snively, G., & Corsiglia, J. (2001). Discovering indigenous science: Implications for science education. *Science Education,* 85(1), 6–34.

Solano-Flores, G., Lara, J., Sexton, U., & Navarrete, C. (2001). *Testing English language learners: A sampler of student responses to science and mathematics test items.* Washington, DC: Council of Chief State School Officers.

Solano-Flores, G., & Nelson-Barber, S. (2001). On the cultural validity of science assessments. *Journal of Research in Science Teaching*, 38(5), 553–573.

Solano-Flores, G., & Trumbull, E. (2003). Examining language in context: The need for new research and practice paradigms in the testing of English-language learners. *Educational Researcher*, 32(2), 3–13.

Songer, N. B., Lee, H.-S., & Kam, R. (2002). Technology-rich inquiry science in urban classrooms: What are the barriers to inquiry pedagogy? *Journal of Research in Science Teaching*, 39(2), 128–150.

Songer, N. B., Lee, H.-S., & McDonald, S. (2003). Research towards an expanded understanding of inquiry science beyond one idealized standard. *Science Education*, 87(4), 490–516.

Southerland, S. A. (2000). Epistemic universalism and the shortcomings of curricular multicultural science education. *Science & Education*, 9(3), 289–307.

Southerland, S. A., & Gess-Newsome, J. (1999). Preservice teachers' views of inclusive science teaching as shaped by images of teaching, learning, and knowing. *Science Education*, 83(2), 131–150.

Spillane, J. P., Diamond, J. B., Walker, L. J., Halverson, R., & Jita, L. (2001). Urban school leadership for elementary science instruction: Identifying and activating resources in an undervalued school subject. *Journal of Research in Science Teaching*, 38(8), 918–940.

Stanley, W. B., & Brickhouse, N. (1994). Multiculturalism, universalism, and science education. *Science Education*, 78(4), 387–398.

Stanley, W., & Brickhouse, N. (2001). Teaching sciences: The multicultural question revised. *Science Education*, 85(1), 35–49.

Stoddart, T., Pinal, A., Latzke, M., & Canaday, D. (2002). Integrating inquiry science and language development for English language learners. *Journal of Research in Science Teaching*, 39(8), 664–687.

Sutherland, D., & Dennick, R. (2002). Exploring culture, language and the perception of the nature of science. *International Journal of Science Education*, 24(1), 1–25.

Tate, W. F. (1997). Critical race theory and education: History, theory, and implications. In M. W. Apple (ed.), *Review of Research in Education*. Vol. 22 (pp. 195–247). Washington, DC: American Educational Research Association.

Tate, W. F. (2001). Science education as civil right: Urban schools and opportunity-to-learn considerations. *Journal of Research in Science Teaching*, 38(9), 1015–1028.

Teachers of English to Speakers of Other Languages. (1997). *ESL standards for pre-K–12 students*. Alexandria, VA: TESOL.

Tharp, R., & Gallimore, R. (1988). *Rousing minds to life: Teaching, learning, and schooling in social context*. Cambridge: Cambridge University Press.

Thomas, W. P., & Collier, V. P. (1995). *A longitudinal analysis of programs serving language minority students*. Washington, DC: National Clearinghouse on Bilingual Education.

Thomas, W. P., & Collier, V. P. (2002). *A national study of school effectiveness for language minority students' long-term academic achievement*. Washington, DC: Center for Applied Linguistics/Center for Research on Education, Diversity, and Excellence.

Tobin, K. (2000). Becoming an urban science educator. *Research in Science Education*, 30(1), 89–106.

Tobin, K., & McRobbie, C. J. (1996). Significance of limited English proficiency and cultural capital to the performance in science of Chinese-Australians. *Journal of Research in Science Teaching, 33*(3), 265–282.

Tobin, K., Roth, W., & Zimmerman, A. (2001). Learning to teach science in urban schools. *Journal of Research in Science Teaching, 38*(8), 941–964.

Tobin, K., Seiler, G., & Smith, M. W. (1999). Educating science teachers for the sociocultural diversity of urban schools. *Research in Science Education, 29*(1), 69–88.

Torres, H. N., & Zeidler, D. L. (2002). The effects of English language proficiency and scientific reasoning skills on the acquisition of science content knowledge by Hispanic English language learners and native English language speaking students. *Electronic Journal of Science Education, 6*(3) http://unr.edu/homepage/crowther/ejset/. Accessed April 5, 2003.

Valli, L. (1995). The dilemma of race: Learning to be color blind and color conscious. *Journal of Teacher Education, 46*(2), 120–129.

Vélez-Ibáñez, C. G., & Greenberg, J. B. (1992). Formation and transformation of funds of knowledge among U.S.-Mexican households. *Anthropology and Education Quarterly, 23*(4), 313–335.

Villegas, A. M., & Lucas, T. (2002). Preparing culturally responsive teachers: Rethinking the curriculum. *Journal of Teacher Education, 53*(1), 20–32.

Waldrip, B. G., & Taylor, P. C. (1999a). Standards for cultural contextualization of interpretive research: A Melanesian case. *International Journal of Science Education, 21*(3), 249–260.

Waldrip, B. G., & Taylor, P. C. (1999b). Permeability of students' worldviews to their school views in a non-Western developing country. *Journal of Research in Science Teaching, 36*(3), 289–303.

Warren, B., Ballenger, C., Ogonowski, M., Rosebery, A., & Hudicourt-Barnes, J. (2001). Rethinking diversity in learning science: The logic of everyday language. *Journal of Research in Science Teaching, 38*(5), 529–552.

Warren, B., & Rosebery, A. S. (1995). Equity in the future tense: Redefining relationships among teachers, students, and science in linguistic minority classroom. In W. G. Secada, E. Fennema, & L. B. Adajian (eds.), *New directions for equity in mathematics education* (pp. 298–328). New York: Cambridge University Press.

Warren, B., & Rosebery, A. S. (1996). "This question is just too, too easy!" Students' perspectives on accountability in science. In L. Schauble & R. Glaser (eds.), *Innovations in learning new environments for education* (pp. 97–125). Mahwah, NJ: Lawrence Erlbaum.

Warren, B., Rosebery, A. S., & Conant, F. (1994). Discourse and social practice: Learning science in language minority classrooms. In D. Spencer (ed.), *Adult biliteracy in the United States* (pp. 191–210). Washington, DC: Center for Applied Linguistics and Delta Systems Co.

Westby, C., Dezale, J., Fradd, S. H., & Lee, O. (1999). Learning to do science: Influences of language and culture. *Communication Disorders Quarterly, 21*(1), 50–64.

Wideen, M. F., O'Shea, T., Pye, I., & Ivany, G. (1997). High-stakes testing and the teaching of science. *Canadian Journal of Education, 22*(4), 428–444.

Wiley, T. G., & Wright, W. E. (2004). Against the undertow: Language-minority education policy and politics in the "age of accountability." *Educational Policy, 18*(1), 142–168.

Wilson, S. M., & Berne, J. (1999). Teacher learning and the acquisition of professional knowledge: An examination of research on contemporary professional development. In A. Iran-Nejad & P. D. Pearson (eds.), *Review of Research in Education* (pp. 173–209). Washington, DC: American Educational Research Association.

Wong-Fillmore, L., & Snow, C. (2002). *What teachers need to know about language.* Washington DC: Center for Applied Linguistics.

Yerrick, R. K., & Hoving, T. J. (2003). One foot on the dock and one foot on the boat: Differences among preservice science teachers' interpretations of field-based science methods in culturally diverse contexts. *Science Education*, 87(3), 390–418.

Index

ability grouping or tracking, 124–126, 127, 131, 137, 151
academic success, scarcity of research on, 158
accountability and education policy, 27–30, 134–136, 150, 153, 159–160
ACTs. *See* American College Test
Adler, J.
Advanced Placement (AP) Exams
    science outcomes for nonmainstream students, 16
    systemic reform and number of students taking, 131, 134
Africa
    code-switching, 86, 88–89
    worldview of students from, 38–42
African American students
    code-switching, 87–88
    communication and interaction patterns, 43
    computer technology, use of, 65
    differential treatment of, 73
    science education as sociopolitical process, 49, 81, 82
    science outcomes and achievement gaps generally, 15
    scientific reasoning and argumentation, 47
    worldview of, 38
Aikenhead, G. S., 62, 70
Akatugba, A. H., 40
Allen, N. J., 42
alternative assessment formats, 94–96
alternative views of science, 2, 23–26
Amaral, O. M., 85–86, 90, 120, 149
American College Test (ACT)

science outcomes for nonmainstream students, 16
systemic reform and number of students taking, 131, 134
American Indian students. *See* Native American students
Anderson, C. W., 48, 127
APs. *See* Advanced Placement (AP) Exams
argumentation skills, 46–48
Ashmann, S., 127
Asian American students
    culturally relevant science instruction, 141
    as "model minority," 13
    science outcomes and achievement gaps generally, 15
assessment, 92. *See also* standardized testing
    accommodations for ELL and SD students, 96–97
    and accountability, 27–30, 134–136, 150, 153, 159–160
    authentic or performance, 117, 153, 155
    bias in, 151
    consistency of, 98
    cultural factors in, 93–96, 98, 99–100
    distinguishing science knowledge from English proficiency, 97–98
    effects on curriculum and instruction, 136
    of ELLs, 94, 96–99, 100, 151
    formats used in, 94–96
    performance or authentic, 117, 153, 155
    policy context of, 134–136, 152, 159–160
    scarcity of instruments for science achievement testing, 153
attitudes toward science and achievement gaps, 17–18

"equity metric" for systemic reform, 128–129
ERIC database used for research synthesis, 163
ESL/ESOL instruction
  effect on teaching subject areas to ELLs
  inclusion of students in regular classrooms, 131
  mastery of language and, 100
  research synthesis including references to, 165–166
  science outcomes, determining, 21
ethnicity. *See* race/ethnicity, *and also specific groups, e.g., Hispanic students*
ethnoscience, 24
explicit instruction of content, 76–77, 156

family and community, 138. *See also* cultural beliefs and practices
  connection between science education and, 140–142
  educational level of parents, 138–140
  future research agenda on, 160–161
  homeless children, 142–143
  implications of research on, 143–144
  influence of parents on children, 138–140
  influence on science learning of, 138–140
feminist theory, 27, 142
Fishman, B. J., 64, 132–133
4-H community youth program in science, 141
Fradd, S. H., 43, 44, 68, 73, 76, 77, 98, 118
Full Option Science Series (FOSS), 67, 71
Fusco, D., 143
future research agenda, 154
  academic success as opposed to failure, 158
  accountability and assessment as education policy, 159–160
  bilingualism, 158
  computer technology, 157
  ELLs, 156, 160
  family and community, 160–161
  inquiry-based science instruction, 157
  multidisciplinary approaches, need for, 155
  prior knowledge and experience of students, 156–158, 160–161
  science outcomes, measurement of, 154–155
  student diversity, 155–156
  teacher education, 158–159

Gamoran, A., 127
García, J., 60

gardening as culturally relevant science curriculum, 141, 143
Gardner, C. M., 17, 18
Garrison, L., 85–86, 90, 120, 149
Geier, R., 64
gender issues, 11, 27, 34, 59, 61, 155
George, Y. S., 21
Gess-Newsome, J., 105
Gilbert, A., 81, 82
Giroux, H., 45. *See also* border crossing
Glasson, G. E., 106
Gomez, L. M., 141
Grandy, J., 36
Gray, B., 40

Haberman, M., 103
Haertel, E., 95
Haitian students. *See also* Chèche Konnen Project
  communication and interaction patterns, 43
  cultural congruence in science instruction, 73
  curriculum and curriculum materials, 68
  scientific reasoning and argumentation, 47
Halverson, R., 126, 127
Hamilton, L. S., 34
Hammond, L., 84, 140
Hampton, E., 67, 71
Hart, J. E., 67–68, 93–94, 113, 114, 119, 120, 149
Hausafus, C. O., 139
Hayes, M. T., 81
Heath, S. B., 80
Hedin, B. A., 36
Heikkinen, M. W., 61
Hewson, P. W., 129
high-stakes assessment. *See* assessment
highly effective schools, characteristics of, 136
Hill, O. W., 36
Hill, S., 35, 139
Hispanic students. *See also* Chèche Konnen Project; ELLs
  communication and interaction patterns, 43
  cultural congruence in science instruction, 73, 77
  curriculum and curriculum materials, 66–69
  inquiry-based instruction for, 83
  science outcomes, 15